Rethinking Transgender Identities

This volume explores the diversity and complexity of transgender people's experiences and demonstrates that gendered bodies are constructed through different social, cultural and economic networks and through different spaces and places.

Rethinking Transgender Identities brings together original research in the form of interviews, participatory methods, surveys, cultural texts and insightful commentary. The contributing scholars and activists are located in Aotearoa New Zealand, Brazil, Canada, Catalan, China, Japan, Scotland, Spain, and the United States. The collection explores the relationship between transgender identities and politics, lived realities, strategies, mobilizations, age, ethnicity, activisms and communities across different spatial scales and times.

Taken together, the chapters extend current research and provide an authoritative state-of-the-art review of current research, which will appeal to scholars and graduate students working within the fields of sociology, gender studies, sexuality and queer studies, family studies, media and cultural studies, psychology, health, law, criminology, politics and human geography.

Petra L. Doan is Professor of Urban and Regional Planning at Florida State University and a leading international researcher on the topic of trans geographies and planning.

Lynda Johnston is Professor of Geography at the University of Waikato, Aotearoa New Zealand, and an expert in feminist, embodied and queer geographies and more specifically in geographies of gender, sexualities, tourism and social justice.

Rethinking Transgender Identities

Reflections from Around the Globe

Edited by Petra L. Doan and Lynda Johnston

Routledge
Taylor & Francis Group

LONDON AND NEW YORK

First published 2022
by Routledge
4 Park Square, Milton Park, Abingdon, Oxon OX14 4RN

and by Routledge
605 Third Avenue, New York, NY 10158

Routledge is an imprint of the Taylor & Francis Group, an informa business

© 2022 selection and editorial matter, Petra L. Doan and Lynda Johnston; individual chapters, the contributors

British Library Cataloguing-in-Publication Data
A catalogue record for this book is available from the British Library

Library of Congress Cataloging-in-Publication Data
A catalog record for this book has been requested

ISBN: 978-1-472-48601-1 (hbk)
ISBN: 978-1-032-12637-1 (pbk)
ISBN: 978-1-315-61370-3 (ebk)

DOI: 10.4324/9781315613703

Typeset in Bembo
by Apex CoVantage, LLC

Contents

Figures

About the Editors

Petra L. Doan is Professor at Florida State University and conducts research on planning and the LGBTQ community. In addition to numerous journal articles, her most recent edited book *Planning and LGBTQ Communities: The Need for Inclusive Queer Space* was published by Routledge in 2015.

Lynda Johnston is Professor of Geography and Assistant Vice Chancellor Sustainability at the University of Waikato. Lynda conducts research and activism on the challenges and spatial dimensions of inequalities and is one of the editors of the *Routledge Handbook of Gender and Feminist Geographies* (2020).

Contributors

Vinícius Cabral is Researcher of the Group of Territorial Studies at State University of Ponta Grossa. His research focuses on the relationship between space and human rights, with special attention on trans-sexualities. He is a member of the *Renascer* (Reborn) non-governmental organization (NGO), which works in support of human rights of LGBTQIA+ groups.

Débora Lee Comasseto Machado is trans-activist and coordinator of the *Renascer* (Reborn) NGO, which works in support of citizenship and the human rights of LGBTQIA+ groups in Ponta Grossa, Paraná, Brazil. Currently, she develops actions for HIV/AIDS prevention and health promotion with sex workers.

S.P.F. Dale is an independent scholar whose research looks at x-gender, non-binary identity and transgender issues in Japan. Dale received a PhD in Global Studies from Sophia University (Japan) and has taught at Hitotsubashi University, Sophia University as well as other institutions in Tokyo.

Miqqi Alicia Gilbert (AKA Michael A. Gilbert) is Professor Emeritus of Philosophy at York University. S/he has many scholarly articles in Argumentation Theory and Gender Theory. Miqqi Alicia is a life-long cross-dresser, written up in numerous newspapers and magazines, and has appeared in many television and radio interviews.

Tommy Hamilton (they/them/he/him) is a tauiwi, trans masculine, endosex, able bodied and neuro typical person. Tommy is a narrative therapist working across rainbow and mainstream services and SOGIESC community development and sustainability in Aotearoa/New Zealand.

Aiwan Liao is a non-binary identified translator and activist of gender diversity. They began working for Chinese queer movement in 2008, before they earned a master's degree in anthropology and had some unsuccessful attempts for a PhD. Their transformation of self and society goes on.

Sonny Nordmarken is Assistant Professor of Sociology at Georgia State University. His research examines the relational processes that maintain complex inequalities in the lives of transgender people and the strategies social actors use to resist these operations of power.

Marcio Jose Ornat is Professor of Geography in the graduate program at the State University of Ponta Grossa and Vice-coordinator of the Group of Territorial Studies at the same university. His research is focused on the relationships between space, gender, sexualities and religion.

And Pasley is in the midst of completing their doctorate in education at the University of Auckland. Their research broadly focuses on gender, sex, sexuality and disability, employing a posthuman lens. Their work is dedicated to attending to the inequalities that haunt our present as a means of fostering a more just world to come.

Maria Rodó-Zárate is Serra Húnter Fellow at the University of Barcelona. Her research focuses on the study of social inequalities from an intersectional, spatial and emotional perspective applied to issues such as the right to the city, gender-based violence or LGBTI phobia.

Rae Rosenberg is Lecturer in Human Geography for the School of Geo-Sciences at the University of Edinburgh. He is a cultural and urban geographer interested in the everyday resistances of multiple-marginalized LGBTQ2+ peoples and communities to structures of racism, colonialism, homonationalism and transphobia.

Chloe Schwenke is President of the Center for Values in International Development, and Adjunct Professor at the School of Public Policy, University of Maryland, USA. Her career includes project experience in over 40 countries. She served under President Obama, working on LGBTQ+ human rights issues globally.

Joseli Maria Silva is Professor of geography in the postgraduate program at the State University of Ponta Grossa. She is Coordinator of the Group of Territorial Studies at the same university. Her research is focused on the relationship between space, gender and sexualities, with special attention on trans-sexualities.

Amets Suess Schwend holds a PhD in Social Anthropology, MA in Art Therapy and MA/BA in Sociology, and works in research and teaching at the Andalusian School of Public Health, Granada, Spain and collaborates as PhD supervisor at the University of Granada. Their recent publications focus on trans and intersex studies, human rights, depathologization and research epistemologies, methodologies and ethics.

Avery Brooks Tompkins is Sociologist at Transylvania University in Lexington, Kentucky. His teaching and research interests include the body, inequalities, community, identity and digital technologies.

Jaimie Veale is Senior Lecturer in Psychology at the University of Waikato. She is the Principal Investigator of Counting Ourselves: The Aotearoa New Zealand Trans and Non-binary Health Survey. Jaimie is on the Board of Directors of the World Professional Association for Transgender Health (WPATH).

Acknowledgments

This book has been a long time in the making. We acknowledge all involved in getting it to publication. We thank all of the contributors for not only taking the time to write for this book but also writing about deeply important and challenging issues. We thank the team at Taylor and Francis, particularly Nonita Saha (editorial assistant) and Faye Leerink (editor).

1 Introduction

Under, Beside and Beyond the Transgender Umbrella

Petra L. Doan and Lynda Johnston

The field of transgender studies has emerged in the context of an ongoing debate over terminology within trans communities. Struggles over names include debates over whether to use transvestite or cross-dresser, whether to spell transsexual with one s or two, whether transgender is a broad umbrella term or is simply a term to describe non-operative transsexuals (á la Virginia Prince), whether a person is transgendered or transgender (e.g., can transgender be an adjective, identity noun or both?) and whether transgender is an umbrella term with broad or narrow scope. To understand these tensions over names, it is useful to remember that – as Kathy Charmaz (2006) notes – naming is an action imbued with power.

> Names carry weight, whether light or heavy. Names provide ways of knowing – and being. Names construct and reify human bonds and social divisions. We attach value to some names and dismiss others. . . . Names, then, are rooted in actions and give rise to specific practices.
>
> (306)

This volume seeks to explore an array of trans identities from across the global north and south. The terminology used reflects a broad diversity of expressions unique to these varied places, cultures and experiences. How we name ourselves reflects our history, our geographies and our uniqueness. It is seldom helpful to critique each other's names. In fact, the evolution and many facets of our movement have been marked by too much strife over naming. Perhaps it is time to move beyond the words and feel the place that the words come from.[1]

In this chapter, we consider the multiple terms and politics associated with transgender and how this, in turn, shapes debates in transgender studies. We pay attention to the power of language as an active force, as neatly described by Susan Stryker's (2006a: 254) articulation of transgender as "an umbrella term that refers to all identities or practices that cross over, cut across, move between or otherwise queer socially constructed sex/gender boundaries". In what follows we first trace some of the initial politics of transgender naming and representation before, second, turning to queer trans studies, language, politics and

DOI: 10.4324/9781315613703-1

places. Here we illustrate the way in which transgender studies continue to be an academic discipline, in its own right, as well as showing the way in which trans studies infiltrate and influence an array of other disciplines. Here we reflect on our experiences of the politics of naming within various trans movements and places. Finally, we provide brief summaries of the chapters contained in this book from scholars and activists based in Scotland, the United States, Brazil, Japan, China, Spain, Catalan and Aotearoa New Zealand.

Transgenderist and Alternatives

To understand the evolution of words in trans communities is probably best to start with Virginia Prince (1997), a revered and occasionally reviled elder in the United States, who claimed to have invented the term transgenderist. Her leadership in the community was evident in providing an early place for people suffering from society's inability to understand differently gendered people. Prince founded the Hose and Heels Club in Los Angeles in the 1950s (Stryker 2006b). Later she published several magazines for people wishing to explore what she called "the other side" including *Transvestia*, and *Femme Mirror*. Petra found copies of both these publications as she was exploring her own gender identity as a middle-aged person. Prince seems to be the first person to have used the term "trans" to refer to a range of different identities.

> There are three classes of such 'trans' people, generally called 'transvestites, transgenderists and transexuals'.
>
> (Prince 2005: 42)

She appears to have coined both the terms transgenderism and transgenderist (Ekins and King 2005), defining transgenderist as:

> The second class is a group of which I am a member and about which most of you haven't heard, namely that of the transgenderists. These are people who have adopted the exterior manifestations of the opposite sex on a full-time basis but without any surgical intervention.
>
> (Prince 2005: 43)

Prince, however, also sparked strong criticism for some of her positions. Richard Ekins and Dave King (2005) describe her detractors as follows:

> Despite the huge influence of Prince's writing and activism over almost half a century, towards an acceptance of transvestism as a lifestyle, her approach and philosophy has attracted fierce criticism over the years, both from within and without the transgender community. It has been depicted as homophobic and sexist and has been criticised for its failure to engage with the issues of sexual politics raised by the women's and gay movements.
>
> (12)

Petra's dear friend Holly Boswell was the first person she knew personally who used the term transgender, and it was a wonderful elixir. Boswell (1991) used transgender as a kind of middle ground.

> I realize this term (heretofore vague) also encompasses the entire spectrum: crossdresser to transsexual person. But for the purpose of this article – and for what I hope will be a continuing dialogue – I shall attempt to define transgender as a viable option *between* crossdresser and transsexual person, which also happens to have a firm foundation in the ancient tradition of androgyny.
>
> (29, italics added)

Holly spent much of her life celebrating her transgender identity and helping others to find joy in their gender difference. Holly originated the transgender symbol blending the male and female signs (see Figure 1.1). She regularly hosted groups at her home, the Bodhi Treehouse near Asheville, NC as well as the Kindred Spirits retreats in nearby Hot Springs, North Carolina where she explored the spiritual dimensions of gender difference.

Leslie Feinberg was a self-defined gender warrior who envisioned a transgender movement for people who felt excluded by the gender binary. When Petra first met Les Feinberg at a reading of *Stone Butch Blues* at Rubyfruit Books in Tallahassee, sometime in the mid-1990s, she remembers being absolutely entranced by the passion and power that was evident in every word that was spoken. An early pamphlet by Feinberg (1992) self-identified as transgendered as the following quote from *Transgender Warriors* (1996) makes clear.

> 'You were born female, right?' The reporter asked me for the third time. I nodded patiently. 'So do you identify as female now, or male?' She rolled

Figure 1.1 Transgender symbol created by Holly Boswell

her eyes as I repeated my answer. 'I am transgendered. I was born female, but my masculine gender expression is seen as male. It's not my sex that defines me, and it's not my gender expression. It's the fact that my gender expression appears to be at odds with my sex. Do you understand? It's the social contradiction between the two that defines me.

(101)

This quote clearly had an influence on Petra's own self-identification. In her (2010) article titled the "Tyranny of Gendered Spaces", she describes herself as a transgendered woman. Subsequently she was startled to receive an email from a young trans activist, who informed her that by using the phrase transgendered she had made it impossible for them to promote the Tyranny of Gendered Spaces in their blog. Pauline Park, a noted trans activist in the New York metro area, has had similar experiences when she describes herself as transgendered. Park's (2011 and 2014) blog posts do an effective job deflating the criticisms of transgendered as an incorrect grammatical usage, but do little to silence the outraged voices raising vehement objection to the term transgendered. This level of vitriol is quite puzzling. Just as trans people can choose their own names, why can't they also name their identities? Why this need to censure others within communities? The answers to these questions are, of course, complex and by tracing the changing terminology across place and time we can get a better sense of the politics of individual and collective trans identities.

A Queer Umbrella

A queer theory lens provides some perspective on this conundrum. Queer theory seeks to destabilize the very category of gender. Kate Bornstein (1994) began queering transsexuality and extending transgender theories. Annamarie Jagose (1997) highlights the instability of identity categories, arguing that "queer is an identity category that has no interest in consolidating or even stabilizing itself. . . . [Q]ueer is always an identity under construction" (p. 131). It is the act of renaming an otherwise marginalized identity by deconstructing and destabilizing the concepts that provide a sense of agency and a means to reclaim power.

Queering gender is what Lynda has been doing since realizing, as a young teenager, that narrow, essentialist and binary constructs of masculine and feminine do not map easily onto lived, felt and embodied experiences (Johnston 2019; Johnston and Longhurst 2016). In the 1990s, Lynda completed a Masters and a PhD, both of which are heavily informed by queer, feminist poststructuralist and embodiment theories that positively acknowledge difference. Paying attention to the body and the visceral brings to the fore lived and diverse experiences of gender. Queer theorists who use gender only as a playful category, however, must remember that the very real difficulties experienced by some whose expression of gender difference can lead to often painful consequences. For instance, Sally Hines argues:

> I believe that poststructuralist deconstructions of gender categories are useful in analysing gender diversity, so long as they are attentive to the lived experiences of difference. . . . I have attempted to show that (trans) gender identities are cut through with difference, while the concept of difference itself is contingent upon social, cultural, political, temporal and embodied considerations. This is significant when considering the divergent identity positions and varied subjectivities which fall under the broad umbrella of 'transgender'.
>
> (Hines 2006: 63)

Back in 1992, Feinberg highlighted the importance of naming ourselves, being inclusive of difference, and the wonderful problem of an outdated pamphlet:

> There are other words used to express the wide range of 'gender outlaws': transvestite, transsexuals, drag queens and drag kings, cross-dressers, bull-daggers, stone butches, androgynes, diesel dykes, or berdache – a European colonialist term. We didn't choose these words. They don't fit all of us. It's hard to fight an oppression without a name connoting pride, a language that honors us. In recent years a community has begun to emerge that is sometimes referred to as the gender or transgender community. Within our community is a diverse group of people who define ourselves in many different ways. Transgendered people are demanding the right to choose our own self-definitions. The language used in this pamphlet may quickly become outdated as the gender community coalesces and organizes – a wonderful problem.
>
> (5–6)

For trans people the act of naming and renaming ourselves provides an active effort to overcome oppression and claim a new vision. As such it is a process that is often in flux and is sometimes hard to keep up with. Trans historian Susan Stryker (1998), recognizing this rich diversity, suggests that what is needed is a more inclusive metaphor for the concept of transgender.

> I use transgender not to refer to one particular identity or way of being embodied but rather as an umbrella term for a wide variety of bodily effects that disrupt or denaturalize heteronormatively constructed linkages between an individual's anatomy at birth, a non-consensually assigned gender category, psychical identifications with sexed body images and/or gendered subject positions, and the performance of specifically gendered social, sexual, or kinship functions.
>
> (149)

Yet Megan Davidson (2007) interviewed 90 trans-identified activists to explore the meanings of the transgender umbrella and found many inconsistencies, suggesting that the umbrella concept is somewhat flawed. Identity politics in the trans community are as active as the identity politics that plagued the early gay liberation movement. Issues of assimilation and radical change continue to dominate discussions with some groups longing to be accepted and assimilated and others linking their movement to broader social change. Davidson comes down on the latter position and emphasizes the potential for change.

> The trans movement nicely exemplifies the creation of new social and political understandings and meanings, challenging and changing the boundaries of the U.S. social imaginary in terms of the possibilities for sexed bodies, gender identities, sexualities, and personhood.
>
> (79)

Over the past decade or so, there has been a rapid expansion of gender identities claimed by individuals both claiming shelter under that umbrella and from those who feel excluded from that categorization. This discursive tension throws into question a number of powerful binaries, particularly cisgender/trans (Darwin 2020; Roen 2002). The strains to the umbrella metaphor are clearly illustrated by the findings of the *National Transgender Discrimination Survey* that surveyed 6,546 trans-identified individuals and revealed a diverse range of identities. The survey asked respondents to identify both their gender identity and sex assigned at birth, but they included a flexible question (Q3) so that those whose gender was not listed (GNL) could name their identities. The authors of the study note:

> Q3 garnered 860 written responses to GNL, many of them creative and unique, such as twidget, birl, OtherWise, and transgenderist. The majority of these respondents wrote in genderqueer, or some variation thereof, such as pangender, third gender, or hybrid. Still others chose terms that refer to third gender or genderqueers within specific cultural traditions, such as Two-Spirit (First-Nations), Mahuwahine (Hawaiian), and Aggressive (Black or African American).
>
> (Harrison, Grant, and Herman 2012: 14)

These 860 respondents constituted 13% of the sample providing clear evidence of a profusion of identity categories. Furthermore, 23 respondents refused to acknowledge any gender, 19 individuals suggested that they perceived gender as completely fluid, 16 others claimed a third gender status, and ten people opted for more radical terms such as "gender-fuck" or "radical-fuck".

In a subsequent study, Stachowiak (2016) conducted more detailed qualitative interviews with genderqueer individuals and found that these people experience a difficult struggle to negotiate their own experience of gender in the face of external oppression. Clearly these individuals also struggle with normative gender expectations, but some are not comfortable under a narrowly construed transgender umbrella.

There is a call to interrogate the whiteness of transgender studies as well as decolonize trans knowledge (Stryker and Currah 2014). In Aotearoa New Zealand the identity "takatāpui" is sometimes used as an umbrella term for gender and sexually diverse Māori, the Indigenous peoples of Aotearoa, meaning "intimate companion of the same sex, 'Takatāpui'" (Williams 1971, 147). Ngahuia Te Awekotuku (1991: 37) – Māori scholar and activist – issued a challenge: "we should reconstruct the tradition, reinterpret the oral history of this land so skillfully manipulated by the crusading heterosexism of the missionary ethic". The missionary ethic was also cissexist, refusing to acknowledge the diverse gender identities of South Pacific peoples, for example:

> Some Māori may use the term whakawahine, which literally means to be like a woman, or tangata ira tane, which means to be like a man. For Samoan people, the term fa'afāfine is often used to refer to being like a woman, while fa'atamaloa – or tomboys – is to be like a man. In Tonga, the term fakaleiti is to be like a woman, and in the Cook Islands akava'ine is to be like a woman, and in Fiji vakasalewalewa. These terms are heard in the Pacific region, yet, they are also contested, are performed differently depending on place, and, of course, there are many more gender identities and expressions.
>
> (Johnston 2019: 7)

While some research exists in these Pacific regions – for example: Besnier (1997, 2002, 2004, 2011); Hutchings and Aspin (2007); Kerekere (2017); Schmidt (2001, 2003, 2010, 2016); Te Awekotuku (2001); Tcherkézoff (2014) – much more is needed if we are to be serious about decolonizing transgender scholarship. Recognition of the intersectionality of embodied difference is one way to do this.

Perhaps, then, as Petra argues, it is time to replace the well-worn notion of an umbrella.

> After a strong storm, especially one with high winds, one often finds the tattered remains of blown-out umbrellas strewn along the sidewalks. The results from the National Transgender Discrimination Survey suggest while that it is time to discard this metaphor for queerness, there is no replacement terminology that includes and possibly shelters all the people gathered in the vicinity of this blown-out umbrella.
>
> (Doan 2016: 98)

Contents of This Collection

This book illustrates that sheltering under the umbrella is still important, yet so is the need to problematize the universality of this concept. In seeking a diversity of views, experience and scholarship, we are delighted that authors come from a range of subject positions across both global north and global south communities. The contributing scholars and activists are located in Aotearoa New Zealand; Brazil; Canada; Catalan; China; Japan; Scotland; Spain and the United States.

The identity and lived experiences of being a cross-dresser are often not deemed worthy of shelter under the transgender umbrella, as noted in chapter two. Miqqi Alicia Gilbert (who is based in Canada) offers a cross-dressing definition before critically examining western constructions of gender, clothing, social situations and political views. Importantly Miqqi brings his/her own lived experience to the chapter and urges scholars and activists to adhere to feminist politics. Cross-dressers, Miqqi highlights, are often marginalized by other transgender people, which may keep many cross-dressers in the closet. Yet, there are many cross-dressers who defy the identity they were assigned at birth and move back and forth between genders. This demonstrates the artificiality and fragility of a bi-gender dichotomy and the chapter calls for greater acceptance of trans diversity, both under and beyond the transgender umbrella.

The urgent need of gender-affirming healthcare is a major finding of the survey discussed in Aiwan Liao's chapter "Brother ("*xiong di*") Communities in Mainland China". The identifier "Brother" – trans men, female-assigned non-binary people and gender non-conforming women – communities in Mainland China were surveyed in order to create baseline data about an under-resourced and socially marginalized sub-group of gender minorities. The survey also illustrates that institutional protection and cultural acceptance needs to occur in order to reduce discrimination and marginalization in a range of everyday spaces and places.

The importance of embodiment is the subject of the chapter four by Joseli Maria Silva (from Brazil), Maria Rodó-Zárate (from Catalan) and Marcio Jose Ornat (from Brazil). They consider the aging experiences of Brazilian travesties and transwomen sex workers. Their research shows that while travesties and transwomen sex workers prepare for early death (the average life expectancy of travesties and transwomen sex workers in Brazil is 35 years old), those who survive do not think about, or prepare for, an aging body. Focusing on the intersectionality of age, class, gender, sexuality and place, the narratives provide rich insights into conflicting experiences of rejection, marginalization, and yet new social possibilities of comfort and acceptance.

Chapter eleven in this collection brings us firmly back to the politics of naming. Avery Brooks Tompkins (from the United States) considers cisgender women in trans relationships. Documenting white, middle-class, cis women partners of trans people shows that limits of identities such as "straight", "bisexual", "lesbian", "queer" and "pansexual". These categorical frameworks do not allow for the complexities of gender and sexual bodies and relationships. Avery notes that the lack of language may prevent cis people with trans partners from finding and forming communities.

Social inclusion is the topic of the next chapter by Chloe Schwenke. Focusing on the spatial scale of international development, Chloe (writing from the United States) shows that the underreporting and invisibility of trans population baseline data mean that many sexual and gender minorities are often excluded from gaining basic benefits, dignity and care. International relief and donor funds are scarce and highly competitive; hence, there is an urgent need for research to make visible transgender experiences associated with human rights and support. Chloe asks "how do we measure and evaluate the plight of transgender persons, especially in the Global South?" and outlines the challenges associated with gathering data.

The critical examination of the politics and possibilities of designing and implementing qualitative social scientific research with trans people is the topic of chapter six. Rae Rosenberg (a scholar based in Scotland) discusses his research experiences with incarcerated trans feminine individuals in the United States. Employing critical trans politics throughout the research process, Rae reflects on the emotional labor involved in conducting research that challenges transphobic state violence. A politics of intimacy emerges in and through the connections between Rae and the research participants. Rae argues for a multifaceted politics of care in trans-focused research in order to challenge the necropolitical forces of transphobic state violence.

International and regional human rights agencies are, in some places, including transgender and intersex identities and needs into their strategic documents. The chapter by Amets Suess Schwend shows how a depathologization and human rights perspective may shine new light onto the relationship between academic and activist research. Amets, who lives and works in Spain, is motivated by the subject position of being both "researched as trans and researching trans" and the experience prompts a critical investigation into epistemological, methodological and ethical aspects in research practice.

Thinking through transnormativities are And Pasley, Tommy Hamilton and Jaimie Veale. They advocate for importance of place-based knowledge from Aotearoa New Zealand and the surrounding Pacific region in order to put the spotlight on the regulations of trans people's lives in and through the healthcare sector. Furthering notions of majoritarian and minoritarianism, Pasley, Hamilton and Veale provide key examples of gender dynamics and the formation of transnormativities. They critically analyze invisibility and hypervisibility in healthcare spaces and urge healthcare providers to collaborate with trans people for better trans care.

The notions of transnormative and cisnormative form the basis of the chapter by Sonny Nordmarken. Drawing on semi-structured interviews with gender minority participants in the San Francisco Bay Area, Sonny illustrates the way in which misgendering actions hold gender minorities accountable to expectations of both dominant, cisnormative, heteronormative ("cishet") and counter-hegemonic, transnormative gender paradigms. Adding to critical geographies of embodiment, the chapter is a powerful reminder that even within trans and queer spaces, transnormative misgendering occurs due to the assumption that gender must be visible and recognizable.

The next chapter considers gender regulation in Japan. S.P.F Dale focuses on the micro-spatial scale of bathrooms and toilets in order to problematize nationalistic gender essentialism. Toilets are places where – as the chapter illustrates – violence occurs against trans and non-binary people. Toilets are regulated on normative and essentialized binary gender which in turn restricts trans people from using them. This chapter shows how in Japan regulations and laws give rise to discursive violence in the form of scaremongering and controlling which toilet trans individuals may use. Drawing together court cases, media accounts and interviews, Dale shows how non-binary people adopt strategies to counter the alienation they feel in public toilets.

Scholars Joseli Maria Silva, Marcio Jose Ornat and Vinícius Cabral are joined by trans-activist Débora Lee Comasseto Machado in the final chapter titled "When a trans is killed, another thousand rise!" Transnecropolitics and resistance in Brazil. Transnecropolitics is a concept developed in the chapter in order to understand the multi-scalar effects of marginalization, violence and ultimately death of travesties and transwomen in Brazil. Not protected by the state, transpeople, the authors argue, are murder victims in Brazil and considered "superfluous beings" by Brazilian societies.

We hope this trans collection provokes and encourages new thinking about what it means to be under, beside or outside of the transgender umbrella. Each chapter in this book revolves around trans and queer studies, yet, all chapters can be read individually or the book may be read from start to finish. We hope that your reading will prompt you to question further some of the assumptions upon which trans language, naming, bodies and knowledge rests.

Note

1 There is an apocryphal story told by John Woolman a Quaker who in 1761 traveled among several Native American tribes in Pennsylvania's Wyoming Valley during the French and Indian Wars to express a message of peace. At one meeting he prayed aloud in the manner of Friends without using an interpreter and Papunehang, a local chief of the Delaware tribe in Wyalusing, who knew very little English, responded to Woolman's message saying "I love to feel where words come from". P. 133, *The Journal and Major Essays of John Woolman* edited by Phillips Moulton. New York: Oxford University Press, 1971.

References

Bornstein, Kate. 1994. *Gender outlaw: On men, women, and the rest of us*. New York: Vintage Books.

Boswell, Holly. 1991. "The transgender alternative." *Chrysalis Quarterly* 1(2): 29–31.

Besnier, Nico. 1997. "Sluts and superwomen: The politics of gender liminality in urban Tonga." *Ethnos: Journal of Anthropology* 62(1–2): 5–31.

Besnier, Nico. 2002. "Transgenderism, locality, and the miss galaxy beauty pageant in Tonga." *American Ethnologist* 29(3): 534–566.

Besnier, Nico. 2004. "The social production of abjection: Desire and silencing among transgender Tongans." *Social Anthropology* 12(3): 301–323.

Besnier, Nico. 2011. *On the edge of the global*. Stanford, CA: Stanford University Press.

Charmaz, Kathy. 2006. "The power of names." *Journal of Contemporary Ethnography* 35(4): 396–399.

Darwin, Helana. 2020. "Challenging the cisgender/ transgender binary nonbinary people and the transgender label." *Gender and Society* 34(3): 357–380. DOI: 10.1177/0891243220912256

Davidson, Megan. 2007. "Seeking refuge under the umbrella: Inclusion, exclusion, and organizing within the category transgender." *Sexuality Research and Social Policy* 4(4): 60.

Doan, Petra L. 2010. "The tyranny of gendered spaces: Living beyond the gender dichotomy." *Gender, Place and Culture* 17: 635–654.

Doan, Petra L. 2016. "To count or not to count: Queering measurement and the transgender community." *Women's Studies Quarterly* 44: 89–110.

Ekins, Richard, and Dave King. 2005. "Virginia prince: Transgender pioneer." *International Journal of Transgenderism* 8(4): 5–15. DOI: 10.1300/J485v08n04_02

Feinberg, Leslie. 1992. *Transgender liberation: A movement whose time has come.* New York: World View Forum.

Feinberg, Leslie. 1996. *Transgender warriors: Making history from Joan of Arc to Dennis Rodman.* Boston: Beacon Press.

Harrison, Jack, Jaime Grant, and Jody L. Herman. 2012. "A gender not listed here: Genderqueers, gender rebels, and otherwise in the National Transgender Discrimination Survey." *LGBTQ Public Policy Journal at the Harvard Kennedy School* 2(1). Available at: https://williamsinstitute.law.ucla.edu/wp-content/uploads/Harrison-Herman-Grant-AGender-Apr-2012.pdf

Hines, Sally. 2006. "What's the difference? Bringing particularity to queer studies of transgender." *Journal of Gender Studies* 15(1): 49–66.

Hutchings, Jessica, and Clive Aspin. 2007. *Sexuality and the stories of indigenous people.* Wellington: Huia Publishers.

Jagose, Annamarie. 1997. *Queer theory: An introduction.* New York: New York University Press.

Johnston, Lynda. 2019. *Transforming gender, sex and place: Gender variant geographies.* London: Routledge.

Johnston, Lynda, and Robyn Longhurst. 2016. "Trans(itional) geographies: Bodies, binaries, places and spaces." In *The Routledge research companion to geographies of sex and sexualities,* edited by G. Brown and K. Browne, 45–54. Oxon and New York: Routledge.

Kerekere, Elizabeth. 2017. *Part of the Whānau: The emergence of Takatāpui identity. He Whāriki Takatāpui.* Doctoral thesis. Wellington Aotearoa: Victoria University of Wellington.

Park, Pauline. 2011. Blog post: GLAAD is wrong on 'transgender' vs. 'transgendered'. Available at: https://paulinepark.com/glaad-is-wrong-on-transgender-vs-transgendered/

Park, Pauline. 2014. Blog post: 5 misunderstandings about the use of transgendered. Available at: https://paulinepark.com/transgender-usage-5-misunderstandings-about-transgendered/

Prince, Virginia. 1997. "Seventy years in the trenches of the gender wars." In *Gender blending,* edited by V. Bullough, B. Bullough, B., and J. Elias. New York: Prometheus Books.

Prince, Virginia. 2005. "The "transcendents" or "trans" people." *International Journal of Transgenderism* 8(4): 39–46. DOI: 10.1300/J485v08n04_07 (Reprinted from Prince, V. 1978. "The 'transcendents' or 'trans" people.' *Transvestia* 16(95): 81–92.)

Roen, Katrina. 2002. "Either/or" and "both/neither": Discursive tensions in transgender politics. *Signs* 27(2): 501–522.

Schmidt, Johanna. 2001. "Redefining Fa'afafine: Western discourses and the construction of transgenderism in Samoa." *Intersections: Gender, History and Culture* (6):1–16, Available at: intersections.anu.edu.au/issue6/schmidt.html [Accessed 24 February 2014].

Schmidt, Johanna. 2003. "Paradise lost? Social change and Fa'afafine in Samoa." *Current Sociology* 51(3–4): 417–432.

Schmidt, Johanna. 2010. *Migrating genders: Westernization, migration, and Samoan Fa'afaine*, Oxford: Ashgate.

Schmidt, Johanna. 2016. "Being 'like a woman': Fa'afāfine and Samoan masculinity." *The Asia Pacific Journal of Anthropology* 17: 287–304.

Stachowiak, Dana M. 2016. "Queering it up, strutting our threads, and baring our souls: Genderqueer individuals negotiating social and felt sense of gender." *Journal of Gender Studies* 26(5): 532–543. DOI: 10.1080/09589236.2016.1150817.

Stryker, Susan. 1998. "The transgender issue: An introduction." *GLQ: A Journal of Lesbian and Gay Studies* 4(2): 145–158.

Stryker, Susan. 2006a. "My words to Victor Frankenstein about the village of Chamonix: Performing transgender rage." In *The transgender studies reader*, edited by S. Stryker and S. Whittle. New York and London: Routledge.

Stryker, Susan. 2006b. "(De)Subjugated knowledges: An introduction to transgender studies." In *The transgender studies reader*, edited by S. Stryker and S. Whittle. New York and London: Routledge.

Stryker, Susan, and Paisley Currah. 2014. "Decolonizing the transgender imaginary: General editors' Introduction." *Transgender Studies Quarterly* 1(3):303–307.

Tcherkézoff, Serge. 2014. "Transgender in Samoa: The cultural production of gender inequality." In *Gender on the edge: Transgender, gay, and other Pacific Islanders*, edited by N. Besnier and K. Alexeyeff, 115–134. Honolulu: University of Hawai'i Press.

Te Awekotuku, Ngahuia. 1991. *Mana Wāhine Māori: Selected writings in Māori women's art, culture and politics*. Auckland: New Women's Press.

Te Awekotuku, Ngahuia. 2001. "Hinemoa: Retelling a famous romance." *Journal of Lesbian Studies* 5(102): 1–11.

Williams, Herbert. 1971. *A dictionary of the Māori language. Seventh Edition*. First Edition 1844. Second Edition 1852. Third Edition 1871. Fourth Edition 1892. Fifth Edition 1917. Sixth Edition 1957. Wellington: Government Print.

2 Defining a Cross-dresser

Miqqi Alicia Gilbert (AKA Michael A. Gilbert)

The Cross-dresser

Definition: A cross-dresser is a person who sometimes wears the clothes of the opposite gender *because they are the clothing of the opposite gender to which the person has been assigned.*

In recent times the western world has become more familiar with transgender people than ever before. Celebrities such as Chas Bono, Kristen Beck and Caitlin Jenner have made people realize that anyone can or might turn out to be transgender. In addition, television shows like *Transparent* and *I Am Cait* have also introduced the subject and made many aware of the existence of transsexual people. This is not really new. It goes back to the sensationalist television drama of Jerry Springer (**Yes**!!! This beautiful young lady, the girl of your dreams, is really a man!!! [Cue screams.]), to the gentler US television hosts Merv Griffin (who legend has it, was himself a cross-dresser,) and also Barbara Walters who seemed genuinely supportive. I am not going to pursue this historical line, I will, however, make one important point.

In all this media celebration, if you will, of all things transgender, the cross-dresser (CD) is completely left behind. The general public believes that "transgender" is equivalent to "transsexual". Indeed, not long ago I was at dinner with a couple and mentioned I was transgender by virtue of being a cross-dresser. They insisted I was not, and that only transsexuals were transgender. (Why that meant that there was no difference between the two words did not seem to worry them.) In this chapter I first examine the definition I have offered, then the role and importance of clothing in our culture and, especially, to the cross-dresser. The last part will be focused on the place of the CD in the larger trans community and the role that a commitment to cross-dressing can play.

First, an examination of the definition.

"A Person"

Cross-dressing is an intentional activity, which is to say that it involves choice and decision. Humans have traditionally been divided into two sexes, male and

DOI: 10.4324/9781315613703-2

female. This tradition focused at the outset on genitalia, so, people with vaginas were females and women, and people with penises were males and men. This simple definition does not work. As you will read elsewhere in this collection, the range of intersexuality precludes any one of a large variety of simple definitions. Whether we go by genitals, hormones, reproductive capacity or chromosomes, there are always serious exceptions. For just one example, a human with Androgen Insensitivity Syndrome, a condition wherein the human body does not respond to masculinizing hormones, results in an apparent female with a vagina who is, by DNA standards, male. The bottom line is that there is no clear way of creating a stable, usable, definition of sex that can be applied to all persons. Still, our western political, social and cultural institutions behave as if the distinction is clear and stable, so as a result, I am going to use female and male as if they were clear categories.

Gender

While we use sex to bifurcate humans into the two most popular categories, it is in reality not sex, but gender that we rely upon daily. Defining gender in a careful and scholarly way would require a book, not a chapter, so to keep things moving along, I will rely on the definitions and ideas derived from some of the first researchers to examine this issue. Kessler and McKenna (1978) and West and Zimmerman (1987) in turn rely heavily on Garfinkel and Goffman (Garfinkel 1967; Goffman 1976; Goffman 1977). In fact, most current researchers still rely on this work of their predecessors (Risman 1998; Risman 2007; Ridgeway and Correll 2004; Lorber and Wayne 2007; Lorber and Moore 2007). But first, the very simplest definition of the sex/gender distinction attributed to Virginia Prince who said, "Sex is what's between your legs, and gender is what's between your ears" (Prince 1976: np). We know from the aforementioned text that being female or male is a complicated and inexact definition. There are a multitude of ways for being female or male, and we are just beginning to understand them. But compared to how to be a woman or a man, being a female or a male is a breeze, and, moreover, the two may be more connected than we realize (Diamond 2008).

We all learn social rules regarding our birth-designated gender. From day one we are conditioned and socialized to follow certain rules and behave in specific ways. These rules derive from the culture in which we are, and, more specifically, from family, religion, school, entertainment, and on and on. Of course, as we mature we may deviate intentionally from our designated role, and that is where it gets interesting. While there is no one way to *do gender*, to put it in West and Zimmerman's and Judith Butler's (1990) terms, to one extent or another, we all still do it, and most of the time we join or can be seen to be allied to an identifiable group. So, there are girly-girls, butch girls, Goths, preppies and a host of others. But virtually all the persons involved in these categories remain recognizable as female or male, and identify as such. These folks, those who are content with their birth-designated gender, are these days referred to as *cisgendered*. The range of gender must not be ignored for two reasons. First, because regardless of the degree of participation it is always there, and secondly because some people do try to avoid gender and it is not easy even for them. Simply put, there is no such thing as an ungendered person – all bodies are gendered with numerous factors contributing to just how this is done (Butler 1990; Gilbert 2011).

Clothing

The most ubiquitous marker of gender is clothing. And, by clothing, I include all the components of adornment people use including makeup, jewelry, facial and body hair as well as the typical items that cover our bodies. I stress this because, especially in current times, there are numerous items of clothing that are "unisex" and might be worn by anyone. One common example is blue jeans, other includes t-shirts, hoodies and running shoes. Even with such common items, however, there are often differences in style, shape, adornment and such-like. Women's jeans often have decorations that will range from the design of the back pockets, the shape of a t-shirt's collar, sparkles on hoodies, or the texture and color of the materials used. Clearly, there are innumerable items of adornment that are in no way ambiguous and clearly identify the wearer as female or male without an instant's hesitation. Skirts, dresses, flowery, lacy or flouncy blouses, makeup, high heels and other gender-specific clothing declare gender and, ipso facto, sex without equivocation. Cross-dressers often complain that women have enormous latitude in the clothing they wear: they can usually dress as masculine as they like without incurring disapprobation. A man, however, who wears clothing obviously meant for women will receive stares and possibly comments.

Consider comfort. A woman mowing her lawn or weeding her garden will most likely wear something on the order of cut-off shorts and a T-shirt. The same thing a guy would wear. But on a hot August day there is nothing more comfortable than a nice loose summer dress, but let that same guy enjoy its lightness and coolness, and firm gender laws are violated. These differences do not come from nowhere, but are rooted in the patriarchal socio-economic power structures in which we must survive. For a man to adopt an indication that he is moving "down the ladder" of power is shocking since no one wants to abjure power. Consider that the summer dress exposes a great deal of the body, allows comparative easy access to genitals and signals such characteristics as frailty, gentleness and leisure: all non-masculine characteristics. Given that we are all gendered, the task at hand on a daily and moment-to-moment basis is to maintain the gender category to which we desire to be assigned. That is no mean feat, and requires constant vigilance.

Because

One interesting fact of our western gendered world is that females, women, can and sometimes do, choose to wear male clothes. This can be for many reasons, as mentioned before, including durability and comfort. I had friends, a heterosexual couple, who only bought men's cotton bikini briefs and shared an underwear drawer. She found them comfortable and far more durable. So why not? Women wear their boyfriend's shirts, sweaters and jackets without anyone remarking on their attire, unless it is to note "how cute" they look. ("Looking cute" is forbidden to males over 13 years of age.) What is interesting about this and what speaks to my point is that females are not typically dressing in male clothing because they want to be taken, felt or be treated as males. They are not wearing male clothing because it's male, but because it's their boyfriends, it's comfortable or convenient. They do not want to be taken as male or be seen to be male. No, they are still very much females. They might, especially if they are lesbian, want to give off a masculine aura, but they still generally want to be seen as women, albeit butch women.

Men *never* wear their girlfriend's clothes with the possible exception of once for sex and once for a crossover party. (As someone once said to me of crossover parties, women don't like them, and men like them too much.) But the CD is different. He (typically) loves to wear women's clothing, *because* they are women's clothing. The CD loves that the clothing makes him feel feminine or, more accurately, how he imagines feminine feels. There is an old tale about a CD who is in a clothing store. There are two tables of identical sweaters. One has a sign, Men's, and sells for $40 apiece. The other says, Women's, and sells for $60 each. He spends the extra $20 and buys a women's because he wants to wear a woman's sweater. This makes the CD essentially different from a transsexual because the transsexual wears woman's clothes because she is a woman. A CD wears women's clothing because they are women's.

I hope the definition I am using is clear. We have people who are assigned, usually at birth, to a sex based on their visible genitals. This puts them in a sex class which is associated with a specific gender – one of only two. Once in that gender, most people naturally follow *to one degree or another* the rules laid down by society. In the case I am dealing with, western or westernized cultures, the CD is a violator of those norms, and this means he is subject to constraints and punishments.

What I want to turn to now, because it is so central to the role and activities of the CD, is the meaning and significance of clothing.

The Politics of Clothing

Social Politics

Clothing is a political issue. It serves a great number of purposes, all of which are social, but many of which are also political. First, I want you to think about what we know, or, more correctly, assume about someone from their clothes. The first thing that comes to mind is gender, is this a woman or a man. But clothes are not always safe gender signals in every context. I work on a university campus, and as often as not, the young women and men there are wearing the same thing – jeans, a T-shirt and runners. When a young woman dresses like that, she can be stating many things, including, to cite just one, that she is not in a romantic space and is focusing on her classes. This is evidenced by seeing those same young women dressed for a social night out. At the other extreme, noting them during final exam week, you will not see any "attraction embellishments" at all.

Of course, clothes establish a lot – clothes speak volumes. Again, on campus, there are identifiable groups of students. One group always wears black and usually has shocking hair colors. Another looks like they just stepped out of a catalog for country wear, and yet another just walked out of a store for the junior elite. Looks vary from a business-like appearance to quite grungy; some attire is ethnic, some quite ordinary and some quite bizarre. But in each case, her personal attire establishes who a person is and who she wants to be. By this I mean, she is establishing her group connections and, if you know the codes, telling the world how to react and what to expect. The diversity of dress, especially on some place like a campus without a dress code, allows you to establish your identity and membership very easily.

Most of the time women dress differently from men. How differently depends on a number of variables. The first is age. Women who are young and fecund generally dress differently from mature women who are partnered. The second variable is occasion. Clothing at work varies, often dramatically, from clothing at a party or celebration. A great number of women in North America wear slacks to work as opposed to skirts or dresses. Similarly, with hair styles. Most working women over 40 seem to have short hair, while single young (especially heterosexual) women almost all have long hair. Finally, a third major variable is personal taste, politics or religion. Some women refuse to display any skin or wear anything but pants, while others prefer to be dressed in a very "feminine" manner at all times. (Culture, of course, has a lot to do with the parameters in which this occurs as well.)

The key is realizing that how we dress expresses how we want to be treated. A woman who is dressed in a provocative manner wants to be noticed; a woman wearing jeans and a loose sweatshirt is trying to minimize her gender membership, generally to minimize male attention. (Note that butch lesbians, many of whom often dress this way, are rejecting the display function assigned to their gender.) If you notice the way western women dress when they are on display, the social politics becomes clear. There is always, among young women, a great deal of skin on display. A dress is standard, often short and low cut, exposing a fair bit of décolletage. The man beside her is covered in cloth from head to toe, with the only exposed skin being his face and hands. The woman, being exposed, is vulnerable and is fortunate that the fully covered male is at her side, ready to protect and cosset. He will open doors, shield her from strangers and otherwise act in a protective, paternal way. The clothing she wears infantilizes and puts her on display while the clothing he wears emphasizes his strength and importance (Goffman, op. cit.) (Corrigan 2008; Lurie 2000; Whisner 1982). That's social politics. One of my favorite and most blatant examples of this difference is demonstrated in entertainment industry awards events. All the men are wearing tuxedos with only minor variations, and all the women, with a few notable exceptions, have exposed skin and very high heels. The difference is astounding.

Clothing then expresses who we are by our choice of styles, colors and image. It also says how much we know by showing that we are aware of what to wear, when to wear it and how to carry it off. Finally, clothing, and our ability to wear it, shows what we are worth. The guy in the $3,000 Armani suit declares his worth when formal, just as he does when informally wearing a cashmere sweater and Gucci loafers. Wearing the right thing, at the right time, and wearing it well, establishes your place in any number of hierarchies. This is the case for men, but it is infinitely more so for women.

The Cross-dresser

Understanding the cross-dresser is helped by understanding the traditional roles implied and enhanced by women's clothing. This changes with culture, context and time but maintains some consistency with twenty to twenty-first-century western ideologies. As per the previous section, the most relevant of these include the following:

- being alluring and sexual;
- being feminine in a traditional sense of:
    - helpless
    - weak
    - fragile

There are two points to be noticed about these. First, they do not apply to men. Men can be "hot" and even sexy, but not alluring or attractive in a womanly way; and men must not be fragile and helpless. Men must be strong, reliable and confident. Secondly, these characteristics are, by and large, just those that feminism has been fighting against for 150 years. Yet men who cross-dress revel in just these feelings that women's attire encourages. (I have more to say about cross-dressing and feminism later.) So one major factor in cross-dressing is a flight from masculinity (Kaufman 1993). For many men constantly being "A Man" is stressful, and it appears to them that women have it easier. Mostly, when they were girls, women were not the ones with the economic, social and cultural resources needed for feeding the family, protecting them and keeping everyone safe. Even today, with the plethora of single-mother families and double-income families, men often feel that "bread winner" responsibility more keenly. There are often difficulties with these responsibilities such as questions of job security, income, and professional or occupational status. It's as if no matter how much you accomplish, it's never enough. At a social gathering, years ago I heard one woman tell another that she was going to stop working. She'd been working for ten years and wanted a break. A male friend was right by me, and we looked at each other in shock. Neither of us could imagine a man ever saying that, if for no other reason than he would not have a husband to support him. The stress to perform, to support, to provide is constant. It also applies to sex: Men cannot fake an orgasm. It becomes immediately apparent when a man is not aroused, or not sufficiently aroused. There is a reason why this situation often develops into "performance anxiety". It really is about performance.

The attraction to women's clothing typically begins very early around puberty. There's no question but that a large component is sexual. Anyone who denies this is fibbing or self-deluded, especially in the early days of CD exploration. Almost all CDs begin by exploring lingerie, usually panties. They are soft, sensual and forbidden. By putting them on the CD is "getting into her panties", and the activity carries great feelings of arousal leading to masturbation. Again, in the early days, masturbation is followed by shame and fear of discovery. The young boy knows he is not supposed to be attracted and aroused by female lingerie, but there it is. He was, and thoughts of the experience linger and recur. When he first puts the panties on he will feel "girly" and feminine, soft and delicate. He's not allowed to wear pretty things, things that are soft, lacy with pastel colors. He's envious. Envious of the luxury surrounding girls, of their softness, of the fact that they don't have to be good at sports, be strong, be fast, wrestle and be tough. As the CD ages a lot of this will, hopefully, change (cf. The Committed Cross-dresser). But for now, the adolescent CD is living in a miasma of arousal, envy and shame.

CD Facts

There are a couple of central facts about a CD that have a great impact. The first is that he came to cross-dressing around puberty. He may have felt feminine yearnings earlier, but likely began to eroticize women's clothing more or less around the time he discovered masturbation. This is important, because, in terms of his understanding of womanness, it means he lost out on an enormous era of socialization, when he was not paying attention to what it means to be a female. (This is also true of the so-called secondary transsexual, or, as I prefer, adult-onset transsexual.) So, when, sometime later, he begins playing with outfits, his taste will go to the erotic and exotic since he has never had the usual girl-to-girl social limitations placed upon him. Girls learn from and teach their girl peers as well as their mothers, and the CD had none.

Once the CD is beyond adolescence he will typically dress more fully. Panties did it when he was beginning but now there's more to cross-dressing. One important factor in a cross-dresser's choice of clothing is the amount of time he gets to dress. A typical CD may get to dress once a month, and if that's all the opportunity you get, you sure don't want to put on a set of sweats. In fact, the more often someone gets to dress, the likelier it is that they will experiment with casual and comfortable clothing. Gals, my preferred term for CDs en femme, who come to the trans event *Fantasia Fair* (see https://fanfair.info/) for a full week, for example, almost always end up in slacks or jeans at one point or another. With a time frame like that, you get to explore more options than cocktail outfits, and can begin to learn what appropriate dress means. There has also been some change in this regard, change I would call progress. Many years ago, when I attended my first Fair, a good half of the first timers trekked to the orientation wearing heels, which is not an easy feat on Provincetown's streets. The past few years, there have been hardly any not wearing flats. This change reflects the way CDs are learning from each other by investigating the web, and how they are learning to care more about being appropriate.

Fantasia Fair and events like it have had an enormous impact on me and my cross-dressing. The ability to interact with other CDs, and even more importantly with cisgendered people, those who are content in their natally assigned gender, has enabled me to grow and explore my woman-self. As I argued (see Gilbert 2011), the essence of existence is interaction. I believe this is what Butler intended in her famous quote, "Persons only become intelligible through becoming gendered in conformity with recognizable standards of gender intelligibility" (1990). Through interaction we learn how to behave and how to be who we want to be. Fantasia Fair, because it is not isolated and confined to trans folk, but mixed into a village of cisgendered people, means that he could interact and observe, and try to fill in the socialization gaps from one's cross-gendered childhood.

I well remember my very first time arriving in Provincetown, MA to attend the Fair. I went to my inn where I was warmly welcomed by the brand new owners. I changed into a skirt and top, and went for a walk on a beautiful sun-shiny day. I was so overcome by the sense of freedom and joy that walking in the sunshine en femme gave me that I had to sit on a bench and collect myself. The week only improved from there. I attended workshops on how to sound feminine, how to walk and sit. I found by the end of my time there that some practices I had adopted became natural and, in fact, have stayed with me. Now, whenever I am en femme these habits are natural and not forced.

Much more on this can be found in the FanFair Stories (Gilbert Various) where my adventures in Provincetown are detailed. Of course, one joy in Provincetown (Ptown) is that no one fusses about bathrooms, the great trans bugaboo. In Ptown, you use whatever washroom you want and certainly the one you are presenting for. "Urinary segregation" (Bornstein 1994) requires that every human make a public announcement as to what sex category she or he identifies with. Entering the skirted room or the trouser room declares your identity, and woe betide the person who makes a mistake. The cross-dresser is particularly caught in this dreadful dilemma as his genitalia fit one room, but presentation of self another. The solution, of course, is to realize that no trans person has ever attacked a cis-person in a washroom (though the reverse is not true) and get over it. But the binary is owed its due.

The Committed Cross-dresser

I am a committed cross-dresser. The term "committed" encompasses, for me, two essential items. The first concerns one's self-identity, the second an approach to what one is doing that includes a mature thoughtfulness and a reflective view of gender roles. Unfortunately, both items often do not come to the individual cross-dresser until later in life (if ever) though for various reasons, that is, happily, changing. The Internet is having an enormous impact on the entire transgender world. A young person who begins cross-dressing only has to search the web to understand that there are groups and resources, ways to learn and understand. This was not always the case.

Prior to the Internet the heterosexual CD spent a great deal of his life experiencing shame and isolation. For a long time, perhaps almost forever, he has been convinced that he is the only man in the world who is compelled to wear women's clothing. This is not something you share with people. It is a practice that is ridiculed and mocked, and those who do it are sissies and fags and, above all, not real men. The isolation brought on by shame and guilt is frequently so extreme that often there is not one other person in the world who knows about his compulsion.

My personal road to commitment began when my first wife announced to me that I was a transvestite, something she had garnered from her therapist. My reaction was to be nonplussed. On the one hand, I felt categorized and somewhat medicalized – I was in a category, a box, neatly labeled. But on the other hand, it meant I could, like the good academic I was, trot off to the library and read all about myself. I took out all the books my university had on the subject, which if I recall correctly was about three, and found I was far from alone. My beginning of self-awareness came through an external stimulus, but others come from various avenues ranging from being advised by a therapist to stumbling across an Internet site.

Once I can say that I *am* a cross-dresser, that I am *glad* I am a cross-dresser, then I have made a major step in accepting myself for who and what I am, and that is a big part of being *any* kind of person including a transgendered person. The society in which the cross-dresser lives wants him to deny his woman-self, to repress his femininity, to pretend that the feelings, urges and compulsions he feels are not really there. This society is broad and includes as component voices not only right-wing Christian fundamentalists but, sometimes, gays who are embarrassed by boys who are sissies, and lesbians who are affronted by caricatures of real women. This is even true of some transsexuals who occasionally treat CDs as annoying little sisters tagging along and spoiling their grown-up time, or worse, as poseurs who water down and degrade the reality of the transsexual journey.

Being a committed cross-dresser involves an acceptance of myself, and a willingness to acknowledge to myself that I am a cross-dresser, will always crossdress, and that, thank you very much, I'm quite happy to be a cross-dresser. Indeed, we can take this even further. The committed cross-dresser comes to see himself as someone who chooses to break the gender rules, who is living, publicly or secretly, beyond the gender laws. In some ways, the cross-dresser is the ultimate gender outlaw. After all, s/he goes from one gender to another, often without passing or even worrying about it. The committed cross-dresser is multi-gendered, or, at least, works at it and tries to involve the crucial aspects, from a personalized point of view, of more than one gender.

A Note on Feminism

An extremely important part of being a committed cross-dresser is being a feminist. How can a cross-dresser be a feminist? First, he can be a feminist in the same way that any other man can. But a CD who is willing to reflect on the realities as well as the fantasies of womanhood can have insights that are not available to many men. These insights range from an intimate awareness of the discomfort of high heels to a shared sense of vulnerability, fear and danger when out alone at night. When the maturing CD comes to appreciate more aspects of his womanness than fun and sexy clothing, then his feminist identity can grow. It's one thing to enjoy wearing high heels, it's quite another to have to wear them all the time. Yes, there are women who like to get "all dolled up", and the CD who wants to is doing nothing wrong. But his feminist instincts should make him pause and think about the very expression. "Dolls" are playthings; they don't have minds or independent choices. They are objects not subjects.

Another arena of feminist self-consciousness concerns the ideas of marginalization and vulnerability: the fact of not being considered, of being left on the side without a voice, without respect and without the ability to make a difference. This has been the reality for women and trans people for a very long time. CDs who are out and about are aware of feeling vulnerable, of potentially being the object of violence and hate. Realizing that women experience this all the time is another step in the CD's feminist journey. In addition, it is important for the committed CD to pay attention to the social dynamics of woman-man interactions. The CD can learn from first-hand experience that many of the issues raised by feminists are real and occur all the time. Being ignored, being treated condescendingly, not having your opinions considered and having men get credit for an idea you had proposed moments before are all real. There are benefits as well. I have, for example, learned that I do not always have to be in charge, but can step back and let others make the decision. I can distinguish between contexts in which my input is important and those in which I can step back.

The simplest way to describe the metamorphosis the CD must undergo to achieve maturity is to realize that while the grass may appear greener on the other side, it is still rife with weeds, worms, bugs, mosquitoes and parts that are tough to deal with. Trying to understand the reality of womanhood rather than just the fun bits is moving further along the road.

Talking about "the reality of womanhood" raises another philosophical issue important to some feminists. Viz., can any man understand what it means to feel like a woman? The philosopher Thomas Nagel raises this issue in a general way in his classic essay, "What Is It Like to Be a Bat?" (1974). I talk about learning more about my "woman-self", but do I ever know that I feel the way a woman feels? Of course, this applies more generally to other people: can I ever know what a Frenchman feels like? A fisherman? A cowboy? If we look at the way a method actor proceeds in learning a part, then we can understand what is involved: it is an attempt to immerse oneself in the relevant culture, values and habits (Gilbert 2001).

My "woman-self" exists and is tutored as a result of paying careful attention to the habits, values and customs of the women around me. She is tutored by my reading, and, yes, by the cultural institutions of the entertainment industry which I definitely take with a grain of salt. It is not easy learning another culture. You have to pay attention to everything from body movement to voice intonation, and end up incorporating them in a natural way. My most cherished times have been when I was spending time with a cis-woman, and at the end was told something like, "I have to tell you, that I felt like I was really talking with a woman". This affirmed for me that I had gone beyond the clothes and entered and allowed my woman-self to be and to thrive.

There are some feminists, a minority, labeled "TERFs" – Trans-Exclusionary Radical Feminists – who insist that no one not born a woman can be a woman. This flies in the face of the difficulties of defining "female" and "male", the reality of the social construction of gender and the reality of innumerable trans people.

Just a Cross-dresser

The trans community in which we all exist is an extremely diverse one. It includes FtMs, MtFs, transsexuals, cross-dressers, drag queens and kings, butches, femmes, intersexes, gender benders and a host of others both pure and in combination. The politicization of the community has led, in recent years, to an increased awareness of our needs and existence in the eyes of the public and various governmental and bureaucratic agencies. As we all know, the awareness that a community exists, that it has a place within the larger society, is a crucial step toward obtaining rights, privileges and respect.

Many of the advances that have been made have been a direct result of activism on the part of the community itself. Everything from urging the inclusion of "T" in LGB organizations to demonstrating at the trial of Brandon Teena's murderers is a step toward recognition and normalization. The existence of the Internet and its vast resources for bringing together disparate and geographically far-flung groups and individuals, not to mention its ability to offer solidarity and anonymity at the same time, has had a major impact on our ability to organize and marshal our forces and energy. In no small part, the introduction of the concept that there is a "trans" community that, though diverse in many ways, nonetheless has a commonality of interest, has enabled a broad support base for many issues. Organizations that had not previously been in contact or seen themselves as part of a larger movement or context now share goals and interests.

The cross-dresser, however, is frequently not a highly respected person within the TG world. His reputation is of someone who has a sexual urge toward women's clothing, and whose experiences typically began early but were primarily masturbatory, that is, a form of fetishistic paraphilia. In addition, he is generally deeply closeted, and if he does come out at all, it is usually to attend restricted gatherings such as club functions or events. At these meetings he will most likely meet and interact primarily with other cross-dressers.

The CD's isolation and separation puts him in a separate class from other trans folk. The transsexual, especially the child-identified transsexual, has the opportunity to inculcate feminine socialization through stealthy observation and selection, though even here it will rarely be complete. The cross-dresser, on the other hand, is normally so conflicted about his gender confusion that the overwhelming shame, guilt and confusion leads to self-repression and a need to distance oneself from feminine identification. Whereas the young female cross-dresser can get away with a "tomboy" identification, the male cross-dresser has no such youthful place to hide (Gilbert 2001).

The average transsexual is much less likely to have the complex erotic and emotional relationship to clothing that the cross-dresser does, especially if the TS is a so-called primary or child-identified TS. The difference is important: The cross-dresser wears women's clothing because they are women's clothes, while the TS wears women's clothing because she is a woman. The reverse is true for the FtM situation, and results in a completely different relationship. Since the child-identified TS has often inculcated cross-gender socialization, the clothes are not exciting but natural. Nor is this to suggest that a TS cannot get excited or aroused or feel erotic as a result of certain clothing, but this is true of all men and women regardless of their birth nature or socially certified gender status. But the chances of seeing an MtF TS hanging about in loose sweat pants or leggings and an oversize shirt are far greater than for a cross-dresser.

All of this sometimes results in the TS viewing the CD in a derisory light where the CD is considered at best a dilettante and at worst a sex-obsessed fetishist who smears the good name of transgenderism. But for most in the community, the CD is, often unconsciously, sort of like an annoying little sister who wants to play with you and your friends but lacks the sophistication and maturity necessary. Yes, she needs to be around sometimes, you are after all related in some way or other, but her eagerness, lack of *savior faire* and inability to measure up make her an embarrassment rather than a friend. She uses too much makeup, wears dreadful clothes, the wrong shoes (usually high heels), always sports those absurd long fingernails, overacts, emphasizes breasts she does not really have and expects to be taken seriously! For goodness sake, most of them don't even know the first thing about feminism, let alone being a woman.

Unfortunately, this attitude permeates, often unconsciously, numerous organizations, projects and undertakings. It is not unusual for cross-dressers not to be invited or involved in political, artistic and scholarly events *not primarily organized by and for cross-dressers*, and when they are there is, not infrequently, a subtle form of marginalization. When I traveled from Toronto, Canada to Oxford, England, for the 3rd International Congress on Gender and Sexuality in September of 1998 I was presenting a paper on socialization. I was not placed in a session with other scholars talking on that subject, but in a session with papers on petticoat punishment and cross-dressing in the theatre. In other words, I was classed not by the subject of my essay, but by my being a cross-dresser. I was, after all, *just a cross-dresser*.

In fairness, the cross-dresser has not always made things easier. There are organizations, for example, that exclude transsexuals from membership for reasons which range from the members' discomfort about homosexuality to concern about SO's ("Significant Others": wives and partners) fears of slippage into transsexualism. In other words, rather than educate their members and their families about transsexuality, it's easier to remain exclusive and discriminatory. This can result in transsexuals not having as strong an organizing base as they might if CDs were included, and often being isolated due to insufficient resources. This attitude on the part of some cross-dressers underscores the idea that CDs are not serious about their gender theorizing, and are not reflective about their role as gender outlaws and their place in the wider TG community.

Ultimately, it is the community as a whole that suffers from this divisiveness. With widely divergent groups that can, but do not always help each other, we hamstring ourselves. Cross-dressers need to realize that they are transgendered, and that, hello, no one really does know if you will wake up one day and want to go full time or sign up for SRS. Adult-onset transsexualism does happen, and it happens to cross-dressers who were certain all their lives that they were just having fun with their "hobby" (McCloskey 1999). Transsexuals also have to understand that there are many CDs who are changing their self-definition, for whom the terminology of "cross-dresser" is becoming too narrow or restrictive. Many of us are making great efforts to grow toward a TG ideology that goes well beyond any reasonable conception of mere fetishism. Many cross-dressers are highly reflective about who and what they are and how that relates to femininity, womanness and the concept of gender.

CDs fear being considered transsexuals who want operations and are gay. TSs fear being considered fetishists who only want to get off. FtMs fear being lost among groups that have been long organized. Intersexuals fear being misunderstood and classed as gender dysphoric. Everyone has fears, far more than I can list here; but isn't it interesting that we all have them? Being gender diverse within a rigidly gender bipolar culture is, after all, terrifying. Maybe if we come together and learn to be less afraid of each other, we'll also learn to be less afraid of the outside world.

References

Bornstein, K. (1994) *Gender outlaw: On men, women, and the rest of us.* New York: Vintage Books.

Butler, J. (1990) *Gender trouble: Feminism and the subversion of identity.* New York, NY: Routledge.

Corrigan, P. (2008) *The dressed society: Clothing, the body and some meanings of the world. Theory, culture & society.* London and Thousand Oaks, CA: SAGE Publications.

Diamond, M. (2009). Clinical implications of organization-activation effects of hormones *Hormones and Behavior, 55,* 621–632.

Garfinkel, H. (1967) *Studies in ethnomethodology.* Englewood Cliffs, NJ: Prentice-Hall.

Gilbert, M. A. (2001) 'A sometime woman: Gender choice and cross-socialization', in Haynes, F. and McKenna, T. (eds.) *Unseen genders: Beyond the binaries.* New York: Peter Lang, pp. 41–50.

Gilbert, M. A. (2011) 'Esse es interagere: To exist is to interact, or, there is no life in the closet', *American Philosophical Association Central Meeting*, Minneapolis, MN.

Gilbert, M. A. (Various) *FanFair stories.* Transgender Tapestry. Available at: https://yorku.academia.edu/magilbert/Fantasia-Fair-Writings.

Goffman, E. (1976) 'Gender display', *Studies in the Anthropology of Visual Communications*, 3, pp. 69–77.

Goffman, E. (1977) 'The arrangement between the sexes', *Theory and Society*, 4(3), pp. 301–331.

Kaufman, M. (1993) *Cracking the armour.* Toronto: Viking Books.

Kessler, S. J. and McKenna, W. (1978) *Gender: An ethnomethodological approach.* Chicago, IL: University of Chicago Press.

Lorber, J. and Moore, L. J. (2007) *Gendered bodies: Feminist perspectives. Gendered bodies: Feminist perspectives.* New York: Oxford University Press.

Lorber, J. and Wayne, L. (2007) 'Breaking the bowls: Degendering and feminist change (2005)', *Atlantis*, 31(2), pp. 110–111.

Lurie, A. (2000) *The language of clothes.* 1st Owl books edn. New York: Henry Holt.

McCloskey, D. N. (1999) *Crossing: A memoir.* Chicago, IL: University of Chicago Press.

Nagel, T. (1974) 'What is it like to be a bat?', *Philosophical Review*, 83(4), pp. 435–450.

Prince, V. C. (1976) *Understanding cross dressing.* Los Angeles: Chevalier.

Ridgeway, C. L. and Correll, S. J. (2004) 'Unpacking the gender system: A theoretical perspective on gender beliefs and social relations', *Gender & Society*, 18(4), pp. 510–531.

Risman, B. J. (1998) *Gender vertigo: American families in transition.* New Haven, CT: Yale University Press.

Risman, B. J. (2007) 'The declining significance of gender?', *Contemporary Sociology: A Journal of Reviews*, 36(3), pp. 238–239.

West, C. and Zimmerman, D. H. (1987) 'Doing gender', *Gender and Society*, 1(2), pp. 125–151.

Whisner, M. (1982) 'Gender-specific clothing regulations: A study in patriarchy', *Harvard Women's Law Journal*, 5.

3 Brother ("*xiong di*") Communities in Mainland China

Aiwan Liao

This chapter is about a research project on Brother Communities in Mainland China. "Brother" here refers to trans men, female-assigned non-binary people and gender non-conforming women. The research, based on a survey that collected 241 responses from across Mainland China, aims to fill a gap of baseline data about this most under-resourced, loosely organized and socially marginalized sub-group of gender minorities. Other researchers are expected to carry out further inquiries based on this chapter, community leaders and rights advocates are welcome to use this chapter to inform their work, and readers in general are encouraged to learn about the ways in which "brother" communities are discriminated and marginalized in everyday Mainland China spaces and places.

Research Background

First, I explain core concepts and geographical boundaries of the research. Next, background factors specific to the place of the research are discussed. After binary social conceptions of gender, LGBTIQ community and movement building, and legal status of gender minority people are outlined, medical definitions of transgender will be elaborated.

Gender identity and gender expression are the core concepts of this research. "Gender identity" in this chapter is understood as one's internal sense of belonging to certain gender, may it be women, men or non-binary categories. "Gender expression" in this chapter is understood as one's external behaviors to stylize one's gender, may it be in feminine, masculine or blended forms. These definitions I give here are in line with a recently published UN report (UNDP and CWU 2018).

The adjective of "Chinese" in this chapter applies to Mainland China (interchangeable with "China"), and the usage of "Mainland China" here is slightly different from its official definition. Officially, "Mainland China" means the political territories actually ruled by the People's Republic of China, which consist of 33 provinces/autonomous regions/municipalities/special administrative regions (SAR), including Hong Kong SAR and Macao SAR. In this

DOI: 10.4324/9781315613703-3

chapter, however, "Mainland China" only refers to the 31 provincial districts of the People's Republic of China, not including Hong Kong and Macau.

Currently, the common understanding of gender in China is the differentiation between men and women. Legally, only the two genders of male and female are recognized in official documents such as birth certificate, identity card, household registration and passport. Politically, the gender agenda is synonym of men and women agenda, as in Article 48 of China's Constitution, it states that "Women of the People's Republic of China enjoy equal rights with men in all aspects of political, economic, cultural, social and family life" (National People's Congress 2018), obviously excluding other gender identities from the matter of equality. Social and cultural public discourses on gender are often associated with power relations between husband and wife, or with career opportunities of male and female workers in the labor market. All the facts demonstrate that gender is prevailingly perceived in a binary frame in nowadays Chinese society.

The prevalence of gender binary has far-reaching impacts. As derivatives of gender binary, hetero-normativity and cis-normativity are dominant in ideologies regarding sexuality and gender in contemporary China. Consequently, it is difficult for individuals with non-hetero sexual orientations and/or non-cis gender identities to gain social acceptance. They encounter challenges in all aspects of life when demanding equal treatment with their heterosexual and/or cisgender counterparts (UNDP 2016). Gender binary does not only bring about unequal power relations among people of different sexual orientations or different gender identities, it also creates hierarchy among people of the same sexual orientation and the same gender identity. For instance, according to my community experience, it is common for trans men who have medically transitioned to feel superior to those who have not done so, since it is believed that medical transition makes them more "real" of a man.

While people with non-hetero sexual orientations and/or non-cis gender identities have always existed in Chinese history, self-organizing and civil activism based on sexual and gender identities to achieve social acceptance and institutional recognition are unprecedented until the last decade of the twentieth century, when Mainland China was opened to waves of globalization (UNDP and USAID 2014). It was against such backdrop that sexual and gender minority people became more visible, claiming identities of gay, lesbian, bisexual, transgender, intersex and queer (LGBTIQ) etc. These identities, along with their adaption to local politics and culture, have since then become the major flags under which non-heterosexual and non-cisgender people in today's China come to terms with themselves and "come out" to others.

For the time being, one can find grassroots organizations working for sexual and/or gender minority people in almost every provincial district. These organizations are often more established in the country's eastern part, in metropolises and provincial capitals, while they tend to be less developed or even absent in the central and the western part, in smaller towns and rural areas. Apart from this geographical pattern, gay men's groups are usually more resourced than lesbian groups, and the latter are more resourced than transgender groups (UNDP and USAID 2014). As consequence of the global leadership in sexual and gender rights of the United States, many community leaders and grassroot organizations in China have been influenced by the American model of movement. The American symbols of rainbow flag for gay pride and blue-pink-white flag for trans pride are widely adopted in China, which showcases an example of this influence.

Upon a closer look at the legal status of gender identity and gender expression, laws in Mainland China do not prohibit non-conforming behaviors such as dressing in a style not based upon one's gender assigned at birth ("cross-dressing"), nor have modern Chinese laws ever forbidden gender-affirming measures including surgical and/or biochemical modifications of the body. In terms of legal recognition and protection, among people with non-cis gender identities, only certain types are entitled to legal gender recognition and marriage/adoption rights: those who have had gender reassignment surgeries and are heterosexual as in the sense of their self-claimed gender identity. Independent anti-discrimination laws are still absent in Chinese legal system, and gender identity and/or gender expression are not protected (UNDP and CWU 2018).

Finally, I look at the matter from a medical perspective, mental and psychological health in particular, where non-normative gender identities are closely linked with pathologization. In 2001, the Chinese Classification and Diagnostic Criteria of Mental Disorders-3rd edition (CCMD-3) indicated that "gender identity disorder" is diagnosed by the criteria that this person "persistently and intensely feels painful" being their natal sex, and longs to be the opposite sex ("not for the cultural or social advantages that becoming that sex may bring about"), while this person "insists on" dressing in the clothes of the opposite sex or "prefer"/"crave for" engaging in activities of the opposite sex, and "stubbornly denies" the anatomy of their natal sex (Chinese Medical Association Psychiatric Branch 2001).

Theoretically speaking, two points are remarkable in the diagnostic criteria of "gender identity disorder" in CCMD-3. On one hand, the pathologization of transgender is absolute in the sense that it does not depend on whether the person accepts their non-normative gender identity. Put in other words, even if a person feels proud to be trans and does not seek conversion of their gender identity, they can still be considered mentally ill (while in the diagnostic criteria of homosexual and bisexual in CCMD-3, only those "self-incongruent" gay or bi people, those who do not accept their sexual orientation and seek therapy for conversion, are considered mentally ill). On the other hand, the pathologization of transgender is relative, since "gender identity disorder" here is defined by its persistence, intensity, non-utilitarianism and binary nature. A typical definition of transgender, however, only emphasizes one's gender identity does not align with one's gender assigned at birth, without specifying whether this person fits into gender binary or resents the anatomy of their natal sex. For these reasons, I argue that, instead of saying transgender, as a vast realm of identities and experiences, is currently pathologised in China, the more precise way to describe it is that, certain types of non-cis gender people who meet the diagnostic criteria of "gender identity disorder" may be considered mentally ill.

The stigma of being a "patient" has negative impact upon all people associated with mental illness, not particularly upon trans people. Due to this fact, the attempt to avoid stigma by solely deleting transgender from the list of mental illness will prove to be short-sighted and narrow-minded. It is also noticeable that, in practice, the treatment for transsexuality differs from that for homosexuality and bisexuality: treatment for transsexuality is typically medical interventions to affirm one's gender identity (modification of the body), while treatment for homosexuality and bisexuality is usually psychotherapy intending to alter one's non-hetero sexual orientations (correction of the mind). Considering that the overall logic under which the Chinese medical system operates nowadays is still "the cure of disease" rather than the "care of health", pathologization is indeed the key element enabling trans people who need medical transition to access gender-affirming surgeries and hormones. In light of this, I assert that it would appear thoughtless to simply call for depathologization of transgender. In the meanwhile, the recent progress made by the World Health Organisation in revising the International Classification of Diseases and Related Health Problems – 11th revision (ICD-11) is worthy of following up by Chinese medical system. In this revision published in June, 2018, "Gender identity disorders" under the category of "Mental and behavioural disorders" has been replaced by "Gender incongruence" and moved to the category of "Conditions related to sexual health", achieving a balance between reduction of stigma related to psychiatric diagnosis and access to gender-affirming medical care (WHO 1990, 2018).

To sum up, the prevailing social perception of gender in a binary structure, the fledgling local LGBTIQ movement in a time of globalization, the neither-persecuted-nor-protected legal status of gender minorities, and the medical definition of transgender as a mental disorder treatable by body modifications are the primary background factors with which participants of this research deal with their gender identity and gender expression.

Survey Profile

At the time when this survey was conducted, there has been no other research dedicated to trans men, female-assigned non-binary people and gender nonconforming women in China. In the LGBTIQ spectrum, the development of Chinese transgender communities is nascent compared to that of gays and lesbians so that there was not as much resources for trans groups. It is also common for transgender people in China to feel uncomfortable with the politicalized LGBTIQ movement and the politicalized transgender identity; thus they tend to keep a distance. This attitude can be attributed to the dominance of hetero-normativity and cis-normativity in sexuality/gender-related ideologies, to the close relation of transgender identity with pathologization, which guards the door to gender-affirming healthcare, and to the seemingly tempting prospect of "passing as normal" after medical transition. Apart from this, unlike their male-assigned counterparts, female-assigned gender-diverse people are generally subject to inferior socio-economic conditions because of patriarchal oppressions. Because of the reasons as stated before, these people are largely unseen and unheard, and knowledge about them is a new territory to navigate.

As a tentative first step, a survey was conducted, in the name of Young Tree, which is a volunteer group I formed with a few friends to empower Chinese female-assigned gender-diverse people. From Oct 24, 2016 to Jan 30, 2017, the survey was available online; all participants were voluntary and anonymous. The survey was posted on multiple social media platforms in Mainland China. Additional promotion methods included email lists, personal reminders and small posters exhibited at a hospital renowned for gender-affirming surgery and at several community centers. Existing participants were asked to forward this survey to potential participants they knew, and small amount of cash bonus was given out through lucky draws to encourage participation. Certain settings were made to guarantee validity, including that the same terminal device and the same IP address could only participate once. Upon its closure, the survey had a total number of 241 participants, coming from all the 31 provincial districts of Mainland China except for the Tibet Autonomous Region.

This survey was titled "Survey of Brother Communities in Mainland China", and "brother" is used here as an inclusive concept, encompassing trans men, female-assigned non-binary people and gender non-conforming women. What the three gender identities have in common is that they were all assigned female at birth (though some of them may be born with typical female anatomy, and the others with atypical traits that can be categorized as intersex), and that they are all non-normative in terms of gender identity and/or gender expression (for trans men, they have a binary identity of man; for female-assigned non-binary people, they have non-binary identities that are neither men nor women; and for gender non-conforming women, they have a binary identity of woman, but their gender expression defies the norms imposed on women). What differentiates the three categories is which gender identity they know themselves to be, rather than whether or not they have had/they wish to have medical transition.

I chose "brother" as the umbrella term for all the three gender identities due to the following reasons: "Brother" ("兄弟" in simplified Chinese characters, pronouncing "xiōng dì") is an existing term that most Chinese trans men use to refer to themselves (Chinese trans women, correspondingly, call each other "sister" – "姐妹" in simplified Chinese characters, pronouncing "jiě mèi"). The term "brother" has apparent advantages such as affirming their male identity, creating a sense of solidarity by underlining brotherhood and avoiding exposure of trans identity to outsiders. The way that sometimes "brother" is interpreted in community daily discourses, however, has its limitations, since it may take on a misogynous hint, excluding many experiences such as non-binary gender identities and non-conforming gender expressions. Considering both the advantages and limitations, I chose to still use the term "brother" but added further layers to its connotation. In doing so, the subjectivity of most participants can be respected, while the social construction of gender as being binary can be challenged.

Data collected by this survey are divided into five sections: Demographics, Gender-Affirming Healthcare, Legal Gender Recognition, Gender Identity/Expression and Societal Life, and Rights Claims. I transformed all these data into graphs for easier accessibility, and readers can go to https://pan.baidu.com/s/1kUUaMEz to access all visual representations of the research. In this chapter, I select some of the data to present, focusing on a few key findings with graphs and further analysis.

Demographics

The gender identity of participants shows 78.84% are trans men, 15.35% are female-assigned non-binary people and 5.81% are gender non-conforming women. This means that the sample is largely dominated by trans men.

On natal sex of participants (sex traits they were born with), 89.63% were born with typical female anatomy and 10.37% were born with atypical traits that can be categorized as intersex. The percentage of intersex is much higher than the average rate of intersex individuals among the general population. According to intersex activists and medical experts, the frequencies of intersex traits vary from one in 60 to one in 2,000 (Holmes 2016). Seemingly, data from this survey imply a higher frequency of intersex among people with non-cis gender identities. However, many Chinese trans men I know said that it feels more comfortable to claim themselves to be born with flawed male sex traits, rather than to admit that they were born female. Using intersex status as the justification for their non-cis gender identity can thus be considered as a creative way for them to defend their subjectivity. Besides, I have seen in person that on the medical certificate issued to post-surgery transgender patients, some Chinese doctors chose to declare "disorder of sexual development" as the cause instead of "gender identity disorder". This difference of cause between the diagnosis at admission and the proof at discharge is probably a considerate move of the doctors to minimize the patient's chance to be discriminated against, knowing that this certificate will be shown to the police for application of new ID documents, and that the stigma of physical conditions is less compared to that of mental problems. This practice may also have contributed to the higher-than-average level of intersex among the survey participants.

On age distribution of participants, the top three age groups are 18–29 years (74.69%), 30–39 years (18.26%) and below 18 years (5.81%). Participants who are above 40 years are rare in the survey. This means that the sample is relatively younger than the general population, as compared to the population pyramid of China in 2016 (Population Pyramid 2016).

On urbanization level of participants' locations, 75.10% live in metropolises, 21.16% live in cities and towns and 3.73% live in rural areas. This means that the sample has a much higher composition of urban dwellers than the general population, as compared to urban and rural population of China in 2016 (Statista 2018).

On number of participants by province, provincial districts with ten or more participants are either national capital, or coastal provinces, or inland provinces with better economy, while provincial districts with five or fewer participants are mostly located in the economically less developed zones of the north, the north-west and the south-west of China. The higher number in more developed regions suggests a positive correlation between economic development and expression of non-normative gender identities.

Lastly, on participants' willingness to stay in touch, 66.39% provided their emails to be informed of the research progress and 33.61% declined the invitation. This means that the majority of the participants care about the results and impacts of the research, which further indicates a sense of community and awareness of rights.

Gender-Affirming Healthcare

Before discussing the data collected in the section of gender-affirming healthcare, it is helpful to look at the most updated regulation made by National Health and Family Planning Commission of People's Republic of China regarding gender-affirming surgeries. This regulation, titled "Technical Management Regulation on Gender Reassignment", was publicized in February 2017 to replace a 2009 regulation called "Technical Management Regulation on Sex Change Operations (Trial)".

This regulation clearly states that "the removal and reconstruction of genitals along with mastectomy for female-to-male individuals are the major surgeries of gender reassignment techniques". Before performing these major surgeries, the person to be operated on should provide the following documents:

1 police proof of having no criminal records;
2 psychiatric or mental diagnosis of transsexualism;
3 notary verified written request of the surgery by the person to be operated on;
4 proof that the person to be operated on has notified immediate family members about gender reassignment surgeries.

In addition to these, the person to be operated on should also meet the criteria below:

1 has been requesting for gender reassignment for at least 5 years without hesitation;
2 has been receiving mental or psychiatric treatment for 1 or more years but has no effect;
3 not married;
4 aged 20 years or above and has full civil capacity;
5 has no surgical contraindications.

This regulation also states that only "after genital reconstruction of the reassigned gender can the hospital issue medical proof for the person operated on to facilitate related legal procedures" (National Health and Family Planning Commission 2017).

Though having no absolute control over individual medical institution or healthcare professional, this regulation plays a guiding role in performing gender-affirming surgeries. According to conversations I had with some trans men and a gender reassignment surgeon, this regulation is abided by to varying degrees. For instance, at the hospital where most of Chinese FtM surgeries are performed, requirements of age, unmarried status and no surgical contraindications apply to all stages of surgeries. For chest surgery, notary verified written consent from parents is required. And for surgery stages after chest surgery, police clearance, psychiatric diagnosis and notary verified parental consent are all needed. Other requirements in the Commission's Regulation, however, such as notary verified written request of the surgery by the patient, the years of requesting gender reassignment and the years of receiving mental or psychiatric treatment, are not demanded by this hospital for surgeries of any stage.

Apart from the influence of the Commission's Regulation, another key factor affecting gender-affirming healthcare is that all the expenses, no matter surgical procedures or hormone therapies, are burden of trans people themselves, as the costs are covered by neither public health insurance nor commercial medical insurance.

Based on the aforementioned background information, I will discuss data regarding the prevalence of hormone therapy and surgical procedures, the reasons behind participants' different medical decisions, the situation of how participants utilize hormones and surgeries, along with the availability, competence and friendliness of gender-affirming healthcare.

On the prevalence of hormone therapy, overall, 58.92% of the participants are on testosterone. Trans men have the highest level at 71.05%, while hormone utilization drops drastically among female-assigned non-binary people and gender non-conforming women. The positive correlation between male identity and hormone utilization is worth taking note of. Testosterone can bring tremendous changes to one's appearance, which are essential to ease gender dysphoria and facilitate societal life. Compared to surgeries, the one-time investment of hormone therapy is trivial, and hormones can be self-administrated while one must rely on medical institutions for surgeries. As a result, testosterone can be the most effective and affordable way to affirm one's male identity. Sheltered by its effectiveness and affordability, however, side effects and potential long-term health risks may not be paid enough attention to by those who are eager to transition.

On the prevalence of surgical procedures, the average level of having had all surgeries is 6.22%, that of having had partial surgeries is 21.58% and that of having had no surgery is 72.20%. According to current technical availabilities, the term "all surgeries" in this survey means the completion of the three stages of mastectomy, hysterectomy and penile reconstruction. The term "partial surgeries" means one or more but not all the steps mentioned before. Compared to female-assigned non-binary people and gender non-conforming women, again, trans men have the highest level of all surgeries (7.37%) and partial surgeries (26.84%).

On the combination of hormones and surgeries, for all the participants, it is much more likely to solely apply hormones (33.61%) than solely apply surgeries (2.49%). Next to the effectiveness and affordability of hormone therapy, this may also be attributed to that hormone therapy is believed to be less irreversible than surgical procedures. Moreover, compared to trans men (26.32%), the proportions of not initiating medical transition (neither hormone nor surgery) rise drastically among female-assigned non-binary people (81.08%) and gender non-conforming women (92.86%). This signifies that currently in China, it is quite hard for most trans men to claim their male identity without attempting at least some medical interventions. Social factors (such as medical transition as the precondition of legal gender recognition) as well as personal reasons (like the pressure from gender dysphoria) can both be considered to explain the phenomenon. In the meanwhile, social factors and personal reasons tangle with each other; the demanding threshold of legal gender recognition can aggravate one's gender and body dysphoria.

On reasons behind participants' different medical decisions, as shown in Figure 3.1 and Figure 3.2, among participants who are not on testosterone, the leading reason for not initiating hormone therapy is "Not needing hormone therapy"; among participants who have had no surgery at all, the most prominent reason for not having any surgery is "No financial capacity". When excluding reasons such as not needing hormones/surgeries and only considering objective circumstances, as shown in Figure 3.3, the major hindrance for participants needing more surgeries beyond the surgeries they already had is the dissatisfaction about current techniques; the primary hindrance for participants wanting surgeries to take their first step is that they cannot afford it financially;

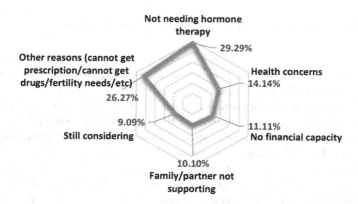

Figure 3.1 Reasons for not taking hormones (*n* = 99)

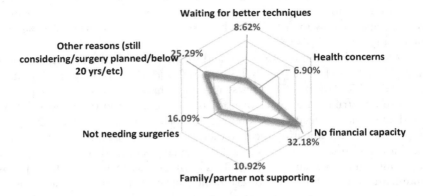

Figure 3.2 Reasons for not having any surgery (*n* = 174)

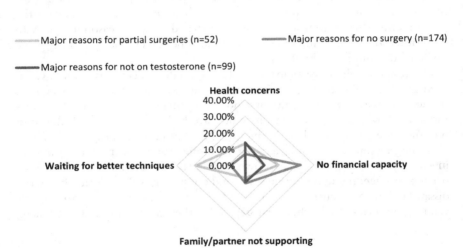

Figure 3.3 Comparison of major reasons hindering body modifications

and the greatest hindrance for participants interested in hormone therapy to be on testosterone is the health risks it involves.

On the situation of how participants take hormones, professional healthcare resources related to cross-sex hormone therapy, including counseling, testing and pharmacy, are extremely inadequate in today's Mainland China, leading to the startling level of unsafe self-medication of hormones. 91.55% of the participants who are on testosterone are taking hormones without medical surveillance, 89.44% are buying hormones from unlicensed vendors and 73.24% are taking hormones irregularly/without laboratory test. To make hormone therapy safer for female-assigned gender-diverse people in China, efforts need to be made in gender-affirming trainings of healthcare professionals, particularly endocrinologists, as well as in the production of cross-sex hormone medications, since most testosterone preparations available now are smuggled drugs sold by unlicensed vendors.

On the situation of how participants have surgeries, there is a low level of involving unqualified healthcare providers (10.45%), as well as a low level of adverse events such as surgical complications (7.46%) or failure (10.45%). This indicates that the participants are much more cautious about surgeries than about hormones, and that compared to hormone services, medical system in Mainland China is doing a better job of providing gender-affirming surgeries. Besides, the fact that only 2.99% of the participants who have had surgeries chose to have surgeries at medical institutions abroad demonstrates the advantages domestic hospitals have in terms of technical competence and service prices. It also reflects the challenges participants interested in having surgeries abroad are facing, including language barriers, higher medical and travel costs and post-surgery legal difficulties such as getting foreign medical proof of gender reassignment recognized by Chinese authorities.

On the availability of gender-affirming healthcare, 86.57% of the participants needing hormone services and 50.93% of the participants needing surgery services reported them hard to access from public hospitals. This signifies that there is an overall deficiency of availability of gender-affirming healthcare in China, and that compared to the gap between surgery needs and surgery availability, the gap between hormone needs and hormone availability is much wider.

On the competence of gender-affirming healthcare, 65.38% of the participants who have tried hormone services and 48.15% of the participants who have tried surgery services are unsatisfied with public hospitals' competence. This signifies that technical competence of gender-affirming services in Chinese healthcare system is generally limited, and that in comparison with surgery services, there is a severe lack of ability in providing hormone services to those in need.

Lastly, among the three indicators of availability, competence and friendliness that are used to measure gender-affirming services in public hospitals, "friendliness" received the lowest level of negative remarks, as only 28.85% of the participants who have tried hormone services and 16% of the participants who have tried surgery services said they were treated unfriendly. At a time when the growing tension between doctors and patients has become a hot topic in China, this remarkable percentage of supportive attitude as reported by the survey participants doubtlessly earned applause for these trans-friendly healthcare professionals. It also reminds community leaders and rights advocates that finding allies among medical workers is a promising direction.

Legal Gender Recognition

I now focus on trans men and female-assigned non-binary people and legal gender recognition. The other gender identity in this survey – gender non-conforming women, who have non-normative gender expression but still self-identify as women – are not likely to be affected by legal gender recognition problems.

Regarding gender marker alteration on official documents, the background information is that under current regulations, sterilizing operations and genital surgeries are both required as preconditions. Data collected show that only 3.16% of trans men and 21.62% of female-assigned non-binary people have no need to change their gender markers, while 69.47% of trans men and 37.84% of female-assigned non-binary people have the need but are not able to change it. In addition to these female-assigned non-binary people who fail to change their gender markers, a further 27.03% of them expressed their dissatisfaction with the either-male-or-female binary options, illustrating the special dilemma of non-binary people who are left outside the narrow boxes of two genders.

Trans men have a much higher rate of smooth gender marker alteration (20%) as compared to female-assigned non-binary people (0%). This rate of successful gender marker change is higher than trans men's level of all surgeries (7.37%), which appears to be contradictory to the national regulation demanding complete medical transition as the precondition of legal gender recognition. However, it is not a secret among Chinese trans men that some of them managed to have their gender markers altered before completing all the surgeries, since they had an acquaintance among local police officials, or they could afford to bribe one. For this reason, the harsh preconditions of legal gender recognition not only deprived many Chinese trans people the right to be recognized and protected by law but also contributed to the corruption of public systems, further marginalizing those underprivileged trans people who have little social/economic capital.

Apart from gender marker on official documents such as identity card and household registration book, which can be changed after having required surgeries, gender is marked on educational certificates in Mainland China, and cannot be altered once the certificate has been issued. This restriction has huge negative impacts on certificate holders who seek further education and apply for jobs, forcing them to either reveal their history of gender transition or give up the opportunity.

In regard to impacts of ID documents on daily life, 70% of trans men and 51.35% of female-assigned non-binary people have encountered problems at airport security check, hotel checking-in, when applying for schools or jobs and so on, due to the ID documents they held did not match their person. At all these moments of "checking", the participants are highly vulnerable in different ways, threatened by involuntary explanation of their "real" gender, or forced body search, or loss of education/career opportunities. Some of the situations could be extremely humiliating, or even putting the participants' personal safety at risk.

Lastly, on attitudes of government staff handling gender marker alteration, 53.13% of the participants who have tried to change their gender markers reported that the police officials handling the procedure had a neutral attitude toward them. Positive attitudes account for 20.31% and negative attitudes take up 26.56%. This suggests that efforts in public awareness raising and in police sensitivity training both need to be enhanced.

Gender Identity/Expression and Societal Life

Everyday experiences – societal life – are shaped by gender identity and expression. Within educational spaces, 56.85% of the participants reported having been treated badly by peer students, 50.21% have been treated badly by teachers, 57.26% felt tired of school or played truant and 7.05% encountered forced transfer or are dropped out of school. Data grouped by gender identity also illustrate that trans men and female-assigned non-binary people have worse school experience than gender non-conforming women. Overall, negative treatment from peers or teachers reduced the participants' school attendance and, in some cases, forced them to drop out, violating their right to education and casting shadows on their future careers.

When it comes to employment, 65% of the participants with work experience have had difficulties in finding a job, 47.50% have had difficulties in getting promotion, 27.45% have received lower wages and 27.27% have received fewer benefits. Data grouped by gender identity also illustrate that trans men's primary challenge in workplace is getting a job; gender non-conforming women regard equal pay as the most prominent issue; as for female-assigned non-binary people, while job hunting is not easy for them either, they actually encounter the least problems once they become employed. On the whole, infringed employment rights leave the participants in an unstable economic situation and constrain their financial capacity to have medical transition, pushing many of them into a vicious circle of no job, no income, no surgery, no change of ID and no job.

On situation of public services and law enforcement, using public toilets/changing rooms is the greatest challenge for all the participants regardless of gender category (88.38% have had problems). The participants also have considerable difficulties in general healthcare (73.44%) and social aid/social security (58.51%), which affects their health conditions and life stability. And the participants are particularly vulnerable when facing law executors (47.30% have encountered their harassment and 41.91% dare not to seek help from them if threatened by outlaws), who may improperly execute their power due to the participants' non-normative gender identity and/or expression. Similar to education, data grouped by gender identity also illustrate that trans men and female-assigned non-binary people have worse experience than gender non-conforming women in all aspects of public services and law enforcement.

Finally, speaking of social attitudes, improper media coverage is the most prominent problem concerned by all the participants (82.57%), while domestic abuse (74.27%), hostility from strangers (60.17%) and refusal of private business owners (26.97%) are respectively the second, the third and the forth severe issue. As for data grouped by gender identity, there is not much difference in the perception of attitudes of strangers and private business owners; however, with family/partner, the experience of trans men and female-assigned non-binary people is again worse than that of gender non-conforming women, and these two gender identities also have higher percentage of spotting discriminatory media reports on gender-diverse people.

Rights Claims

In the last section of the survey the participants were asked about the changes they want to see in respects of laws and policies, medical care, social attitudes and community development. For researchers who are interested in these issues, for community leaders and rights advocates who want to understand what the urgent needs are, and for the general public who have little knowledge about the minority's lives, it is vital to learn about these changes the participants are calling for, and to help realize them.

In the respect of laws and policies, priorities vary greatly among the three gender identities. As shown in Figure 3.4, trans men mostly care about unconditional legal gender recognition and gender change on educational certificates, female-assigned non-binary people mostly care about education discrimination and non-binary gender options, and gender non-conforming women mostly care about education discrimination and domestic violence.

In the respect of medical care, gender identity also plays a crucial role in distinguishing participants' priorities. As shown in Figure 3.5, depathologization is mainly supported by female-assigned non-binary people and gender non-conforming women, who, interestingly, are the gender groups less or not affected by pathologization. Trans men, being the most impacted group, show very limited interest in depathologization, while they are more concerned about insurance coverage and technical competence of gender-affirming healthcare. Female-assigned non-binary people have little complaint about the

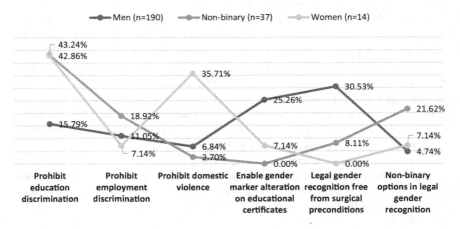

Figure 3.4 Major legal and policy claims – grouped by gender identity

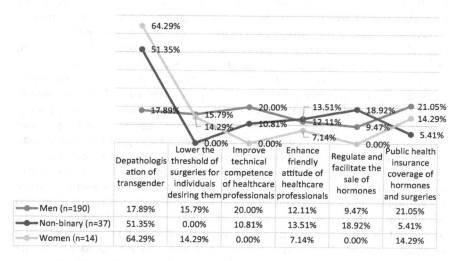

	Depathologis ation of transgender	Lower the threshold of surgeries for individuals desiring them	Improve technical competence of healthcare professionals	Enhance friendly attitude of healthcare professionals	Regulate and facilitate the sale of hormones	Public health insurance coverage of hormones and surgeries
Men (n=190)	17.89%	15.79%	20.00%	12.11%	9.47%	21.05%
Non-binary (n=37)	51.35%	0.00%	10.81%	13.51%	18.92%	5.41%
Women (n=14)	64.29%	14.29%	0.00%	7.14%	0.00%	14.29%

Figure 3.5 Major medical care claims – grouped by gender identity

existing requirements for surgery applicants, while they are more concerned about hormone-related services. And gender non-conforming women's lack of interest in surgical technique and hormone availability echoes with their insignificant need in medical transition.

In the respect of social attitudes, participants' opinions have much in common among different gender identities. Overall, a large percentage of all the participants believe that family/partner support is the most important (43.98%), while proper reporting of mainstream media (17.43%), not discriminated against when openly being trans (16.18%), gender neutral public facilities (9.54%) and acceptance from school environment/workplaces (4.98%) are the rest four items of the top five claims.

And finally, in the respect of community development, opinions again are similar among the three gender identities. Consensus has been reached on raising awareness of gender diversity as being of the highest priority (39.42%), while share medical information (15.77%), fight against internal discrimination (13.69%), promote rights awareness (7.88%) and collaborate with different subgroups of trans people (7.47%) are the rest four items of the top five claims.

Conclusions

The trans community is defined by the diversity it embodies in multiple dimensions. This survey only targets gender-diverse people whose gender assigned at birth is female, and those gender assigned at birth is male are beyond the scope of this study. Although only looking at a small proportion of trans communities, great diversity already exists among the participants. This diversity is reflected in multiple dimensions, including genders they identify with; sex traits they were born with; age; location; willingness to stay in touch; medical interventions and gender marker alteration. These aspects intersect and inter- act with each other, leading to participants' various experiences of education, employment, public services, law enforcement and social attitudes. Diversity is also manifested in participants' opinions regarding laws and policies, medi- cal system, social attitudes and community development. This community is far from united or homogeneous. These inner-community complexities eloquently refute the stereotypes of and prejudices against trans people, and they are important referencing points for gender-diverse people to understand themselves, and for community leaders and rights advocates to create change.

Despite some limitations – such as a sample of younger age and urban dwellers, carried out online (thus I have less control of participants' personal background), and a small number of participants (compared to the estimated population of the entire community) – the results show the need to continue to research brother communities. In particular, gender-affirming healthcare is in urgent demand. This survey dedicates half of its space to gender-affirming healthcare. Regarding the types and degrees of medical interventions of the body, participants of different gender identities made very distinct decisions. Nevertheless, for many of them, the desire to medically transition defines what it means to be transgender. Although the influence of pathologization and the dominance of gender binary are still to be examined, this persistent and profound longing of bodily sovereignty needs to be respected. It is true that medical transition does not solve all the problems trans people face, but even in an ideal social environment with perfect institutional protection and cul- tural acceptance, it is still infeasible for trans individuals who desire medical transition to live well without being able to physically be themselves. This survey suggests that existing gender-affirming services are far from sufficient or satisfying for the participants in need, and the gaps in availability and quality severely undermine the well-being and threaten the lives of trans people. These urgent needs deserve more attention and effort from medical professionals and policy-makers.

Next to the urgent demand of gender-affirming healthcare, the other major finding of the survey is that institutional protection and cultural acceptance need to be strengthened. The second part of the survey centers around insti- tutional recognition and cultural inclusion of female-assigned gender-diverse people. The participants reported different experiences and opinions regarding legal gender recognition, schools and workplaces, public services, law enforce- ment and social attitudes. They also expressed changes they want to see in legal, medical, social and community perspectives. The data collected show that these participants have been challenged in all aspects of societal life due to their non-normative gender identities and expressions. Social structures and cultural practices are not in favor of these participants, depriving them the fair treatment they deserve as equal citizens and human beings. Moreover, trans men, despite having higher level of medical transition, encounter more social challenges than female-assigned non-binary people and gender non-conforming women. This reconfirms the argument that it is impossible to solve all the issues trans people face simply by body modifications. Medical transition (for those in need) must go hand in hand with inclusive and supportive institutional and cultural environments. Only in this way can gender-diverse people be ensured a safe, dignified and fulfilling life.

My heartfelt thanks go to two friends of mine who offered great support to this survey: Di Wang, for your insightful comments, and Bin Xu, for your financial aid. Readers should also be aware that many changes had occurred in the communities since the data were collected. The analysis made in the research thus only represents situations at that time.

References

Chinese Medical Association Psychiatric Branch. 2001. 《中国精神障碍分类与诊断标准第3版（*CCMD-3*）》 [Chinese Classification and Diagnostic Criteria of Mental Disorders-3rd edition]. Chinese Medical Association Psychiatric Branch. Accessed Dec 5, 2021. www.max.book118.com/html/2017/0111/82382877.shtm

Holmes, Morgan, ed. 2016. *Critical Intersex.* New York: Routledge.

NHC (National Health and Family Planning Commission). 2017. 《国家卫生计生委办公厅关于印发造血干细胞移植技术管理规范（*2017年版*）等*15个* "限制临床应用" 医疗技术管理规范和质量控制指标的通知》 [National Health and Family Planning Commission's Release of 15 Medical Technology Regulations and Quality Control Indicators for "Restricted Clinical Application"]. NHC. Accessed Dec 5, 2021. www.nhc.gov.cn/yzygj/s3585/201702/e1b8e0c9b7c841d49c1895ecd475d957.shtml

NPC (National People's Congress). 2018. 《中华人民共和国宪法》 [Constitution of People's Republic of China]. NPC. Accessed Dec 5, 2021. www.npc.gov.cn/npc/c505/201803/e87e5cd7c1ce46ef866f4ec8e2d709ea.shtml

Population Pyramid. 2016. "Population pyramid of China." Population Pyramid. Accessed Dec 5, 2021. www.populationpyramid.net/china/2016/

Statista. 2018. "Urban and rural population of China." Statista. Accessed Dec 5, 2021. www.statista.com/statistics/278566/urban-and-rural-population-of-china/

UNDP (United Nations Development Programme). 2016. *Being LGBTI in China: A National Survey on Social Attitudes towards Sexual Orientation, Gender Identity and Gender Expression.*

UNDP (United Nations Development Programme) and CWU (China Women's University). 2018. *Legal Gender Recognition in China: A Legal and Policy Review.*

UNDP (United Nations Development Programme) and USAID (US Agency for International Development). 2014. *Being LGBT in Asia: China Country Report.*

WHO (World Health Organisation). 1990. *International Classification of Diseases and Related Health Problems (10th revision).*

WHO (World Health Organisation). 2018. *International Classification of Diseases and Related Health Problems (11th revision).*

4 Age, Sexuality and Intersectionalities

Spatial Experiences of Brazilian *Travestis* and Transwomen Aging Process and the Sex Market

Joseli Maria Silva, Maria Rodó-Zárate and Marcio Jose Ornat

This chapter illustrates the intersections of age, place, sex work and *travestis* and transwomen[1] in South Brazil. The research originated from the fact that there are few *travestis* and transwomen who survive the struggle of everyday structural violence. Discrimination, marginalization and violence had impact on their lives, culminating in their premature death (Balzer, Hutta, Adrián, Hyldal, and Stryker 2012; Cabral, Silva and Ornat 2013). There is a need, therefore, to make visible the aging process. The group of participants in this research either worked in the past or still work in prostitution and, according to the *Associação Nacional de Travestis e Transexuais* (ANTRA) (National Association of *Travestis* and Transpeople), constitute the most vulnerable group regarding vulnerability and early and violent deaths. As Antunes (2013) points out, the life expectancy of *travestis* and transwomen in Brazil is 35 years old.[2] This implies that low-income *travestis* and transwomen with poor education who depend on prostitution live daily with the idea of a premature death. This makes them live the present with intensity, with few concerns about their aging process. As a group of gender non-conforming people, their social and economic vulnerability increases in their old age. Unlike the majority of the Brazilian population – who experience a life course in which they expect to age and then die – *travestis* and transwomen expect death but not to experience aging. This group's aging process is specific and has not been extensively explored by gerontologists, who privilege gender normative populations (Ramirez-Valles 2016; Siverskog 2014, 2015) (although some research has focused on gays and lesbians (Clements-Nolle, Marx and Katz 2006; Williams and Freeman 2007; Witten 2009, 2014, 2016)).

In this research, aging is understood as a social construction that is established through the relations in interactions with other groups, as proposed by Hopkins and Pain (2007). Some of the specificities of *travestis* and transwomen aging might be identified in Audrey Hepburn's[3] statement when she was invited to take part in a study and reflect upon the transpeople aging process:

> I don't know. I never think about aging. Do you want to go into therapy with me and make me think about it? I don't think. I am afraid of old age,

DOI: 10.4324/9781315613703-4

this bothers me, wrinkles and the time passing, because I see that in the photos. I cannot see myself like that. I don't know, maybe this is because we coexist with death every day. But now, I'm 40, I see myself as a winner. I look back and see my mates that died when they were 18, 20, 25 and they are not here anymore. Then I am a winner in this sense. But, I also feel that all the burden of my life makes me feel old. When I was thirty, I was already old. But I don't think. But now you make me think, don't you?

(laugh)

Fear of the body aging, the precarious feeling of aging, the presence of death and the lack of perspectives to plan for the future – as reported by Audrey Hepburn – are elements that characterize the specificities of the *travestis* and transwomen aging process. We analyze age, gender, sexuality, social class, place and space through an intersectional lens (Collins 2000; Crenshaw 1991; Davis 2009; McCall 2005; Rodó-Zárate 2013, 2015). In addition, the intersectional experiences present movements along their existence, evidencing a negotiation process that *travestis* and transwomen carry out with the oppression structures around them.

The group of people who took part in the research were self-identified as "old" *travestis* and transwomen, who recall becoming aware of their aging process from a young age,[4] as can be seen in Chart 4.1. Despite the small number of interviewees, the results of this research were discussed and legitimized by a group of 68 *travestis* and transwomen who attended the *XII Encontro Regional Sul de Travestis e Transexuais* (XII *Travesti* and Transsexual people Southern Region Meeting).[5]

The interviews were carried out based on two axes of questions. One of them inquired how aging occurs and the other investigated the transformations of relations between aging, place and space. The group narratives were

Chart 4.1 General profile of the group taking part in the research

Fictitious name	Genre self-identification[1]	Current age	Age/aging perception	Self-definition of economic activity
Audrey Hepburn	Transwoman	42	30	Civil servant/prostitution
Grace Kelly	Transwoman	45	30	Self-employed/prostitution
Marilyn Monroe	Travesti	52	40	Cleaner/prostitution
Sophia Loren	Travesti	43	36	NGO/prostitution
Doris Day	Travesti	62	27	SUS[2] pensioner/prostitution
Brigitte Bardot	Travesti	53	40	SUS pensioner/prostitution

1 The self-identification of sexual orientation is diverse. One of them identifies herself as bisexual, two affirm to be homosexual and the remaining ones state to be heterosexual. However, both groups, the ones that self-identify as heterosexual or homosexual, sexually desire people that they identify as men. Therefore, the declared sexual orientation depends on how they understand themselves in the sex, genre and desire normative structure in the Brazilian context which goes beyond the binary idea of homo/hetero based simply on the genitals. Those who are self-defined as *travestis* tend to identify their orientation as "homosexual" and those who call themselves transwomen see themselves as heterosexual. However, it is impossible to establish a pattern.

2 SUS – Public Health System.

analyzed using content and critical discourse analysis (Bardin 2004; Silva and Silva 2016). We adopted a set of procedures that enabled the mediation between the objective of our research and a view of the reality presented by those who took part in the research (Silva and Silva 2016). The themes were presented to the participants who talked about them. Their speech was transcribed and gathered in a single document. After the statistical treatment of the set of narratives, a word net was built based on the frequency of the use of such words and the number of relations between them.[6] Such procedure was essential to destabilize our initial perspectives, since it revealed dimensions that we had not foreseen. While our initial hypotheses were focused on the negative aspects of aging for our participants, the word net formed by their discourse revealed elements that we had not thought about, such as the positive aspects of aging, for example. After that phase, we carried out the mediation between the elements brought by the group and our research question and created a set of categories that guided the qualitative interpretation of their discourse. At that point, we analyzed the meaning of their speech through marks given to the excerpts

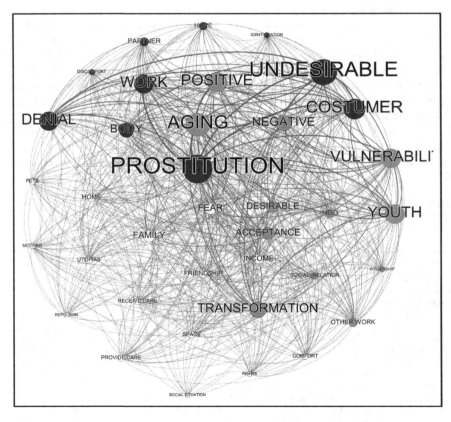

Figure 4.1 General semantic net on the aging process narrated by *travestis* and transwomen in South Brazil

interpreted as units of meaning (categories) so that later on they could be measured and compared. After interpreting the discourse content through the content analysis categories, it was possible to create the net of categories. That semantic net provided a view of the group's discursive structure, evidencing the central and peripheral elements in the set of narratives and showing how those elements were intertwined. Finally, the visual structure of the semantic net allowed us to return to the discursive set and retrieve the most remarkable reports of spatial experiences and the aging process of the participants.

This methodology of discourse analysis allowed the discovery of discursive categories and their higher frequency relations, forming semantic communities that constitute the meanings in the group narratives. Three semantic communities were identified that structured the discourses of the *travestis* and transwomen interviewed, as seen in the different shades and sizes that represent the discourse categories that form the general semantic net.

The semantic communities are explored in two sections. In the first section, we point out that *travestis* and transwomen create the idea of aging in a relational way, considering their vulnerabilities and specific social and economic relations. The second section evidences that the group aging process, despite all difficulties faced in a transphobic society, presents strong resistances, establishing new social, affective and economic links.

"We Have Short Expiration Date": Relational Aging of *Travestis* and Transwomen

> We, *travestis*, have short expiration date. We are different from other people. *Travestis* have an expiration date. If you get old, you get old alone, you can no longer make money. The expiration date is certain for us. It's sad, very sad.
> (Interview with Sophia Loren in Ponta Grossa, 09th May 2015)

Transpeople's needs have been neglected by both the LGBTI community political actions, mainly based on an essentialized male/female binary (Browne 2004, 2006; Browne and Lim 2010; Doan 2007, 2010; Hines 2007; Namaste 2000; Nash 2010). The topic of transpeople and aging has been approached with essentializing understandings of gender. Fredriksen-Goldsen and Muraco (2010) highlight that concerns related to old people in the LGBTI community and access to health institutions, shelters and the public spaces have been directed almost exclusively to gays and lesbians.

Lack of knowledge of transpeople's demands is an aspect criticized by Namaste (2000). She argues that the knowledge produced by transpeople has been mostly concerned with themes such as causes or origin of the transsexuality, identity and performance than the knowledge about their everyday life. According to this author, these are:

> activities of our day-to-day, the fabric of how our bodies are located in, and move through, the world. Although banal, these events merit consideration:

> anything less produces a knowledge of little practical relevance to our lives, reinforcing a world that treats transsexual and transgendered people as inconsequential.
>
> (Namaste 2000, p. 2)

Their life course is made up of daily events experienced in different ways in each phase of life from the childhood, adolescence, youth, adulthood to the old age. Hopkins and Pain (2007, pp. 287–288) point out that

> age and lifecourse stages as socially constructed categories rather than independent variables means that space and place gain significance. People have different access to and experiences of places on the grounds of their age, and spaces associated with certain age groups influence who uses them and how. Further, people actively create and resist particular age identities through their use of space and place.

In this sense, old age, in addition to comprising biological implications such as loss of vitality and the marks produced on the body, is understood as a historical and spatial process of subjectification. Aging for transpeople, who can be marked or not by the conformity to the gender assigned at birth, is made up of specific experiences. However, gender non-conforming people's aging cannot be taken as something homogeneous, since it is not possible to reduce all transpeople's identities to the simple inadequacy between gender and the sex assigned at birth (Siverskog 2015). Transpeople's bodies are inscribed by marks and are positioned in different spaces and times that orient them toward particular movements in the life course (Valentine 2003).

Therefore, the focus of analysis on Brazilian, poor and prostitute *travestis* and transwomen brings about particular forms of life course marked by exclusion and violence, which constitute specificities in their life-stage transitions such as childhood, adolescence, youth, adulthood and old age. Many authors have denounced the precarious life conditions of this group in Brazil. Usually starting around puberty, the abandonment of family, dropout of school and the lack of job opportunities makes prostitution one of the few opportunities for survival (Albuquerque 1995; Baskerville 2012; Benedetti 2005; Boulevard 2013; Lee 2013; Nikaratty 2013; Ornat 2009; Pelúcio 2009; Peres 2005; Riquelme 2013; Santos 2010; Silva 1993, 1996, 2007, 2009).

Their precarious life[7] puts them into a situation of vulnerability and early death risk. Very few *travestis* and transwomen have the privilege of getting old, even though there are some who survive beyond the short lifespan expectation of 35 years of age[8] and go through the aging process.

Aging is, in general, considered a final stage in life, associated to a number of physical and psychological limitations. However, the beginning of old life, as well as its definition, is not something fixed and universal, but variable in space/time and dependent on a series of elements that constitute the experience of getting old (Pain 1997; Hopkins and Pain 2007, 2008). Bodies are marked by the life course, but this is not something uniform. The way the different stages of life are culturally interpreted, as well as the unequal distribution of health technologies, economic and scientific resources, constitute specific aging experiences (Pain, 2001; Pain, Mowl and Talbot 2000).

For the group of Brazilian *travestis* and transwomen investigated, the age close to the thirties is associated to aging, which, in turn, is related to their perception of vulnerability and exclusion, as pointed out by Audrey Hepburn:

> My biggest trauma was when I turned 27 and started to panic about turning 30. It was complete old age in my conception. That is because I knew or was convinced that going from 30 to 100 would occur in the blink of eye. And I was not prepared for that. And the whole life situation, you know, being a transperson. Your life suffers so much prejudice and your opportunities are so few that you cannot even image to go beyond 30 years of age. I remember that I used to say: if I turn 40 it will be too much.
> (Interview with Audrey Hepburn in Curitiba, 19th May 2015)

The perception of early aging reported by Audrey Hepburn shows that age, more than a simple chronological detail, is built in socially established relations (Hockey and James 2003). Brazilian *travestis* and transwomen, that we interviewed, constitute their identities by a strong connection with the sex work market,[9] implying particular ways of aging, which are, in turn, experienced from a particular gender, race, sexuality and social class (Ramirez-Valles 2016).

Figure 4.2 highlights the semantic community, which presents highest frequency and density of relations between the discourse categories. The narratives about aging revealed with highest intensity the correlation between working in the prostitution activity and the body, which, due to aging, is considered refused and undesired by the consumers.

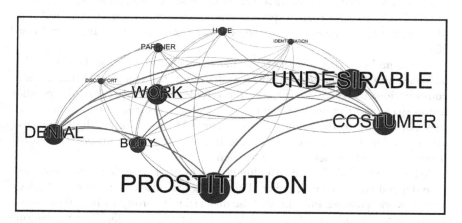

Figure 4.2 Semantic community 1

The body is central in the speech of all *travestis* and transwomen who took part in this research. Their life course is marked by body transformations, either spontaneous or provoked. The end of childhood is determined by the feeling of dissatisfaction with the puberty signals perceived and the struggle to avoid them. Strategies to cheat bodily changes typical of adolescence generate family conflicts and the early moving out from the parents' house, which in turn marks the entrance in the job market to provide for themselves, or their adulthood.

The transition to adulthood is abrupt and narrated by the deepening of affective relations and sociability with other *travestis* and transwomen, as well as by the learning and experimentation of techniques of body transformation, as evidenced by Grace Kelly:

> I left home when I was 13, I was only a child. But then I learned to cater for myself. I lived all phases, I was gay, but I was not satisfied by that. Then I became a *travesti*, and I felt I found myself. Even so, there was something missing. Then, when I was about to turn 18 I injected some silicone. . . . At that time, there was no other option. Nowadays, there is the plastic surgery, prosthesis, at that time, there wasn't! There was not even anaesthetics for the injection. We had to drink like two litres of wine, sit there and the '*bombadeira*'[10] would stick the needle. I remember as if it were today. It would be like sixteen punctures on each breast, and it was like a horse needle, a very big one. It was hard, a kind of torture, you could die there. But it was like that, or you would never be what you wanted to be. Then, we would rather go through it all and see the transition to become the woman we wanted to be.
>
> (Interview with Grace Kelly in Curitiba, 20th May 2015)

Grace Kelly's report highlights the search for the suitable body that would bring psychosocial comfort, from the esthetic pattern instituted in the prostitution space. The sacrifice and risks involved in the body construction are neglected insofar the body is for them a space of concretization of their identity and also of resistance to the gender norms imposed to them (Johnston and Longhurst 2010).

The same body that is rejected in many spaces in the city such as schools, hospitals, churches and legal agencies is desired in the spaces of sexual service consumption, therefore constituting the interdependence between the prostitution space and the body transformation (Silva and Ornat 2014). The more they correspond to the body esthetics desired by their clients, the more service is rendered, and in turn, the more financial resources can be invested to improve the body esthetic desired by them.

"Making the body"[11] alters the life course of poor *travestis* and transwomen who depend on the sexual market. Their bodies are places of identity, esthetics, action, work, pleasure and pain and the material way through which they communicate to the others (Nast and Pile 1998; Valentine 2003). The mediation between *travestis* and transwomen bodies and the symbolic order established, according to Grosz (1994, 1995), creates the conditions of their everyday life throughout the different life stages. Old age is realized in the ritual performances that constitute the prostitution space on the streets. The choice for younger *travestis* and transwomen by the clients is an important element in the constitution of the perception of aging that goes beyond the sexual commercial relations. Being rejected by the clients due to age is a sign of the loss of the femininity that was built with hard work and for many years. It is also a sign of the loss of financial support, as can be seen in Sophia Loren's report:

> Well, nowadays men do not look at me with the same desire as they did when I was younger. I felt this in my daily routine, not receiving the same men's look with the same desire I saw when I was younger. We feel this also in our environment [referring to prostitution] a kind of treatment that is different from the treatment the youngsters receive, the 'nymphets'. They say to us: what is this hideous old bitch doing here? Go home! The youngsters do not worry, because everything is wonderful for them, they are young, and beautiful and they are making money. It was them who showed me that I am old, and also my clients. The clients that used to look for me, stopped to go out with me to go out with them [referring to the younger prostitutes]. And this, of course, makes me face hardship. It is very painful to be an old *travesti*.
>
> (Interview with Sophia Loren in Ponta Grossa, 09th May 2015)

Sophia Loren's report illustrates the connections between discourse categories that make up the semantic community shown in Figure 4.2 and highlights that the aging perception occurs due to the actions permeated by codes, which are specific to that space such as the clients' desire, valuation of youth as attribute of beauty and the tensions between younger and older *travestis* and transwomen. The performance established between the people that occupy the spaces of prostitution is established in the repetition of actions that build the aging process in a relational way (Hopkins and Pain 2007, 2008).

Therefore, the aging process of *travestis* and transwomen who build their existence in connection to the sexual market presents contents and specific life course references, where the body and its transformations are the main markers. When aging is understood as relational these *travestis* and transwomen are considered "old" in this spatial context, and face other challenges that involve the construction of different relations and identities mediated by body transformations.

"Being Called Madam Is Wonderful": Reinventing a Life Possible to Be Lived

In the previous section we addressed the *travestis'* and transwomen's perceptions of getting old from the viewpoint of a social construction of age, deeply related to the vulnerability and dynamics of the sexual market that values young bodies. Johnston and Longhurst (2010) defend that all experiences and subjectification processes are embodied and that binary thinking continues to be a powerful force that structures all bodies, even gender non-conforming bodies. The old and the young individuals form different generations yet are interdependent and relational (Hopkins and Pain 2007, 2008), expressing tensions and collaboration between them.

Those who self-identify as old, narrate experiences which are common in a precarious economic, social, political and cultural context. They say that they are "living on borrowed time" since the Brazilian society denies them the right to get old. Reaching this phase of life is seen by them as "a lucky strike", but, at the same time as the need to face an apprenticeship of living out of the sexual service activity.

Three elements are common in their memories: the violence imposed by the police, mainly during the dictatorship in Brazil[12] (Hutta and Balser 2013); the scapegoating for the HIV virus transmission in the 1980s; and, the esthetics of glamour and exuberance of their bodies. According to them, the female body esthetics of "those times" had beauty references marked by classical and glamorous beauty icons of the Hollywood goddesses. Grace Kelly's testimony is typical of the stories of pride to have survived that period of great political repression in the country, building an authoritative narrative on the younger *travestis* and transwomen:

> In the past, if we went out during the day, we would be arrested. . . . The police would beat us, nowadays there are the human rights for everything, there are laws to protect us, which did not exist before. Then, nowadays I say that they (referring to younger prostitutes) have a sugary life. We didn't. We would be beaten, there was all kind of violence and humiliation. First, the police would beat, then you would be taken to the police station and we would be made to undress and they would use a cold water hose on us. We would spend the night in jail, without having done anything to deserve that, simply for being trans. There was a terrible cop who hated *travestis*, you know? At that time, it was not like today that they (referring to the younger prostitutes) go wearing jeans and trainers. In the past, it was fur coat, tights, very high heels, a lot of makeup, wigs, etc. It was glamorous. The wig was burnt, the coat was torn, the heels were broken and we would be beaten by the police. Then, if they [referring to the young prostitutes] are there pretty and free is because we opened the way for them.
>
> (Interview with Grace Kelly in Curitiba, 20th May 2015)

The semantic community resulting from the *travestis'* and transwomen's narratives, expressed in Figure 4.3, highlights that old age/youth are related one to another, even if in a paradoxical way (Rose 1993). Narratives about aging are compared with their own youth and the current youth. The outcome combines positive and negative elements, in both the evaluation of old age and that of the young.

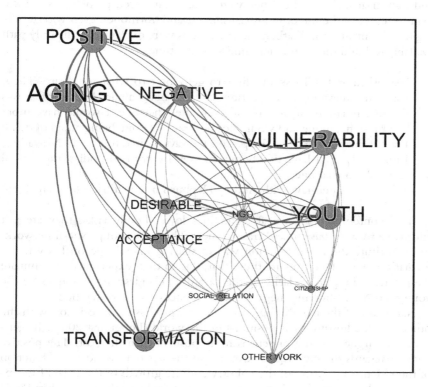

Figure 4.3 Semantic community 2

The recreation of other subjectification ways through aging experiences in a context of extreme violence and exclusion, as in the Brazilian context, shows their capability to fight and resist, since as pointed out by Pile (1996), the bodies are not passive to the socially built meanings, but they resist and reinvent other forms of existence.

Body transformations and the vulnerability are categories, which are narrated in a different way when associated to aging and youth, presenting each of them from both the positive and the negative aspects. "Vulnerability" when associated to youth is linked to violence and death risk, since the youthful trans body is the main target in homicide cases in the country.[13] When associated to the old age, "vulnerability" appears linked to the economic needs, fear of homelessness, food and health care (Browne, Bakshi and Lim 2012). The category "transformation" when associated with "youth" is made up of stories of achievement of female shape through clandestine methods and the emotions of the seduction power over the clients and the lovers. When associated to "old age", the transformation appears in comments about the physical decay and the health complications resulting from the indiscriminate use of hormones and industrial silicone when they were young. However, participants did not express regret of the use of industrial silicone and hormones, even with resulting health impairment. Rather, these decisions were explained as the only path available to build the desired femininity at that time:

> We risked and still risk our lives to get closer to what we desire. Then, I started to have my first injections and was never afraid. My only fear was the first hormone injection and then after that one, I would have more and more hormone and would not be afraid. Despite being afraid of dying I say: 'I died trying to be the person I always desired to be! . . . Nowadays, I joke and say: 'I have two bodies in one. I carry two bodies, the natural and the artificial one.
>
> (Interview with Audrey Hepburn in Curitiba, 19th May 2015)

Even if contemporary surgical techniques lower bodily risks, they are not affordable to poor *travestis* and transwomen. Thus, the group's social networks still constitute the most common way of sharing knowledge on body transformation. The older ones are those that, due to their experience, disseminate the practices, in spite of clarifying the hazards of clandestine techniques for the youngsters, even showing, negative consequences on their own bodies.

Narratives of their body's aging occur in a process of negotiation with the other scales, adjusting their positions in the world (Longhurst 2001). In general, they argue that the present is more positive than the past. The positive content regards the civil rights advances in the last two decades as a function of the redemocratization of the country and the growth of the LGBTI movement, the discovery of medicines and treatment for lethal illnesses such as HIV/ AIDS, as well as the invention of esthetic technologies such as plastic surgery, laser hair removal and transsexual surgery. Such technologies allow the *travestis* and transpeople to transform their bodies with better esthetic results and lower health risks. All the positive advances enjoyed by the younger *travestis* and transwomen are narrated as a result of the fight of the older ones and the construction of a legacy that deserves recognition by the youngsters, as reported by Brigitte Bardot:

> We [referring to the old prostitutes] are leaving our inheritance, everything we achieved to the youngsters, these who are enjoying everything we built. What we went through [referring to police violence at the time of the military dictatorship], there are some [referring to the young prostitutes] who have never gone through anything like that and do not value it.
>
> (Interview with Brigitte Bardot in Ponta Grossa, 08th May 2015)

Intergeneration relationships (Hopkins and Pain 2007, 2008) are established simultaneously through tensions in the sexual market, as previously pointed out, but are also formed through the exchange of care and wisdom between older and younger *travestis* and transpeople.

The awareness of the physical deterioration and loss of sex appeal are simultaneous to the constitution of new sociability and subjectification processes (Grosz 1995). The old *travestis'* and transpeople's bodies become to be interpreted and communicated in a different way from that of the youth time, providing them with novel spatial experiences. There are several testimonies that, in spite of being rejected in the prostitution space, their getting older resulted in greater comfort and acceptance in other spaces which were previously marked by exclusion (Waitt and Gorman-Murray 2007). Doris Day reports enthusiastically:

> It is good to be called madam. I had never imagined that one day I would be called madam, for a person that had always been seen as sexy. . . . Nowadays, being called madam, and more being a respected madam. Wow, excuse me, it is simply wonderful! I think this is luxury!
> (Interview with Doris Day in Ponta Grossa, 12th May 2015)

Old *travesties* connect with places such as school reunions; LGBTI fight/activist spaces and better access to citizenship. This is evidenced in the discourse categories that structure the semantic net shown in Figure 4.3. Those old bodies, now seen as devoid of sex appeal, reconfigure power relations, resistances and meanings, recreating subjectification processes and comprehension of themselves, as shown in Audrey Hepburn's testimony:

> Getting older gave me something very important, which is the acceptance of who I am. Nowadays, I have no problems with myself. I am today a transwoman. But I could say to any person simply that I am a woman. I don't even think of having a surgery [referring to the transsexual surgery], because this is not a priority for me. I can be a woman, even without the surgery.
> (Interview with Audrey Hepburn in Curitiba, 19th May 2015)

The narratives highlight that the construction of an idea of family also goes through transformations with the aging process, as can be seen in Figure 4.4. The narratives evidence that the weakening and sometimes loss of family bonds in the youth are partially reconstructed in the aging time. Conflicts with parents and siblings decrease with time and the *travestis* and transwomen keep links of care with their parents, showing great affection for their mothers. However, the meaning of protection and care is mainly linked to friendship bonds with other *travestis* and transpeople that were built throughout their lives.

Other discourse categories such as fear of the economic future and of not having somebody to rely on in the old age are part of their narratives. These fears, however, are combined with expressions of pride for having reached the old life. Besides that, they identified their homes as places of peace and their pets as elements of affection.

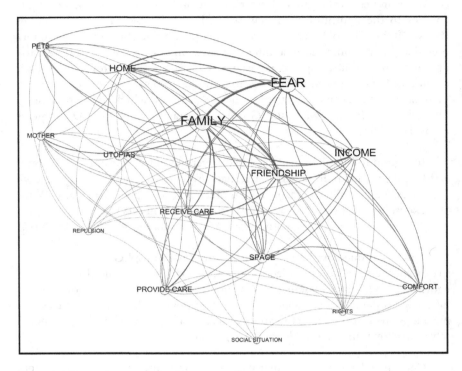

Figure 4.4 Semantic community 3

Finally, despite the recognition of aging as a phase of several physical and economic limitations, *travestis* and transwomen develop strategies, by building new femininities and subjectivities that enable them to go through aging, even if this phase in their life course could not even be imagined by them.

Final Considerations

This research analyzed the way *travestis* and transwomen experiment spatially the process of body aging in Brazil, considering the specific intersectionality of class, age and sexuality. The analysis of narratives produced by the people who took part in this research highlighted that *travestis* and transwomen constitute the idea of getting old in a relational spatial and temporal way. The life course developed by them is marked by the dehumanization of their existences and the naturalization of their premature deaths by the Brazilian society. Aging is considered a privilege for them, since dying prematurely is an evident possibility. The perception of getting old is mainly mediated by the sexual market, strongly marked by the valuation of the young body and the feeling of rejection of their body esthetics in the prostitution space signals their entrance in the old age.

Denial of their presence in the place where they spent most of their life course and constructed the elements of their identity results in economic limitations and new identity challenges. Thus, aging is also socially built by new subjectification processes and different spatial relationships are created by them. Transpeople's aging reduces the notion that their transgressive bodies represent danger to the socially instituted gender order. Therefore, they experiment in a positive way some spatiality that they had been denied in their youth. Finally, the life course developed by *travestis* and transwomen shows that different phases of life are intertwined with identity movements and plural spatial experiences.

Notes

1 We consider the self-identification of the people interviewed. In Brazil it is difficult to find a limit to establish borders of identity between transwomen and *travestis*. Identities are fluid. There are those who stated that they were "still *travestis*", but that after achieving their dream of going through reassignment surgery they would be "transsexuals". Another said that she was currently a transsexual but that she had been a *travesti* before she had undergone surgery. Others say they are transwomen, but they have not had a reassignment surgery. So, to be a *travesti* or transwoman in the Brazilian society involves overcoming the sedentary order of classification structures, involving references with nomadic meanings and narratives.

2 According to the Instituto Brasileiro de Geografia e Estatística (IBGE 2013) (Brazilian Institute for Geography and Statistics) the Brazilian lifespan is 74.9 years old.

3 All names used are fictitious.

4 The Brazilian law considers old a person who is 60 years old or over (LAW N° 10.741, 1st October 2003).

5 The XII Travesti and Transsexual People Southern Region Meeting was held on 18 March 2017 in Rio Grande, coordinated by the Non-governmental Organization "*Construindo Igualdade*" (Building Equality), whose president at the time was Cleonice Araújo.

6 The tool chosen for this part of the research is the software belonging to the statistical packet "R" called "RQDA – an R package for Qualitative Data Analysis". It is an opensource tool presenting the main functionalities required by qualitative data analysis. The platform has an open architecture and stores the data in a data base called SQLITE. The categorization of the transcribed discourse was carried out in RQDA and after that, through simple SQL commands, the data were exported for analysis and visualization of nets in the GEPHI.

7 As understood in Butler's perspective (2004) as a social process of the human beings comprehending transpeople as a menace. Such representation of transpeople constitutes a way to their dehumanization and the justification of their extermination.

8 The Brazilian government does not have official data on the transpeople life expectancy in the country. The Associação Nacional de Travestis e Transexuais (ANTRA) (National Association of Travestis and Transpeople) has confirmed the average of 35 years old presented by Antunes (2013).

9 The National Association of *Travestis* and Transwomen informs that 90% of *travestis* and transwomen survive economically from prostitution in Brazil.

10 "Bombadeira" refers to a person who injects liquid silicone into the body of *travestis* to build a feminine form. She mastered the technique of "building female bodies" for *travestis* by the injection of liquid industrial silicone. This technique is illegal in Brazil. The "bombadeira" is an important person in the relationship networks among the group.

11 "Make the body" is an expression used by the group to refer to the techniques used for body transformation.

58   *Joseli Maria Silva et al.*

12 Military dictatorship in Brazil lasted from 1964 to 1985. In 1989 the Brazilian society voted for the first time to elect a president after the end of the military regime. The redemocratization process was only boosted in the 1990s, but only in 2000 a popular government was elected in Brazil.
13 The National Association of *Travestis* and Transwomen (ANTRA) published in 2017 a map of the murders of *travestis* (84.34 %), transwomen (10.43 %) and transmen (5.21%). The general mean age of those murdered was 27.9 years, and the youngest victim was 16 years old while the oldest was 53. Available at www.google.com/maps/d/viewer?mid=1yMKNg31SYjDAS0N-ZwH1jJ0apFQ&ll=-6.447239100000003%2C-35.412435500000015&z=8. Accessed on: 08/08/2017

References

Albuquerque, Fernanda Farias de. 1995. *A Princesa: a história do travesti brasileiro na Europa escrita por um dos líderes da Brigada Vermelha*. Rio de Janeiro: Nova fronteira.

Antunes, Pedro Paulo Sammarco. 2013. *Travestis envelhecem?* São Paulo: Annablume.

Balzer, Carsten (Carla La Gata); Hutta, Jan Simon; Adrián, Tamara; Hyldal, Peter and Stryker, Susan. 2012. *Transrespect versus transphobia worldwide. A comparative review of the Human Rights Situation of gender-variant/Trans People*. Berlin: Transgender Europe (TGEU).

Bardin, Laurence. 2004. *Análise de Conteúdo* (3rd ed.). Lisboa: Edições 70.

Baskerville, Nelson. 2012. *Luís Antônio – Gabriela*. São Paulo: Nversos.

Benedetti, Marcos Renato. 2005. *Toda feita: o corpo e o gênero das travestis*. Rio de Janeiro: Garamond.

Boulevard, Gláucia. 2013. "Vida de travesti é luta! Luta contra a morte, luta contra o preconceito, luta pela sobrevivência e luta por espaço". In *Geografias malditas: corpos, sexualidades e espaços*, edited by Joseli Maria Silva, Marcio Jose Ornat and Alides Baptista Chimin Junior, 69–81. Ponta Grossa: Todapalavra.

Browne, Kath. 2004. "Genderism and the bathroom problem: (Re)materialising sexed sites, (re)creating sexed bodies". *Gender, Place and Culture* 11, n. 3: 331–346.

Browne, Kath. 2006. "A right geezer bird (man-woman): The sites and sights of 'female' embodiment". *ACME: An International E-Journal for Critical Geographers* 5, n. 2: 121–143.

Browne, Kath and Lim, Jason. 2010. "Trans lives in the 'gay capital of the UK'". *Gender, Place and Culture* 17, n. 5: 615–633.

Browne, Kath; Bakshi, Leela and Jason, Jason. 2012. "Ageing in gay Brighton". In *Lesbian, gay, bisexual and transgender ageing: Biographical approaches for inclusive care and support*, edited by Richard Ward, Lan Rivers and Mike Sutherland, 165–182. London – UK: Jessica Kingsley.

Butler, Judith. 2004. *Precarious life: The powers of mourning and violence*. London and New York: Verso.

Cabral, Vinícius; Silva, Joseli Maria and Ornat, Marcio Jose. 2013. "Espaço e morte nas representações sociais de travestis". In *Geografias malditas: corpos, sexualidades e espaços*, edited by Joseli Maria Silva, Marcio Jose Ornat and Alides Baptista Chimin Junior, 246–275. Ponta Grossa: Todapalavra.

Clements-Nolle, Kristen; Marx, Rani and Katz, Mitchell. 2006. "Attempted suicide among transgender persons". *Journal of Homosexuality* 51, n. 3: 53–69.

Collins, Patricia Hill. 2000. *Black feminist thought: Knowledge, consciousness, and the politics of empowerment*. New York: Routledge.

Crenshaw, Kimberlé Williams. 1991. "Mapping the margins: Intersectionality, identity politics, and violence against women of color". *Stanford Law Review* 43, n. 6: 1241–1299.

Davis, Kathy. 2009. "Intersectionality as buzzword: A sociology of science perspective on what makes a feminist theory successful". *Feminist Theory* 9, n. 1: 67–85.

Doan, Petra L. 2007. "Queers in the American city: Transgendered perceptions of urban space". *Gender, Place and Culture* 14, n. 1: 57–74.

Doan, Petra L. 2010. "The tyranny of gendered spaces – reflections from beyond the gender dichotomy". *Gender, Place and Culture* 17, n. 5: 635–654.

Fredriksen-Goldsen, Karen I. and Ann Muraco. 2010. "Aging and sexual orientation: A 25-year review of the literature". *Research on Aging* 32, n. 3: 372–413.

Grosz, Elizabeth. 1994. *Volatile bodies. Toward a corporeal feminism.* Bloomington: Indiana University Press.

Grosz, Elizabeth. 1995. *Space, time, and perversion. Essays on the politics of bodies.* New York: Routledge.

Hines, Sally. 2007. *Transforming gender: Transgender practices of identity, intimacy and care.* Bristol: Policy Press.

Hockey, Jenny and James, Alison. 2003. *Social identities across the life course.* New York: Palgrave Macmillan.

Hopkins, Peter and Pain, Rachel. 2007. "Geographies of age: Thinking relationally". *Area* 39 n. 3: 287–294.

Hopkins, Peter and Pain, Rachel. 2008. "Is there more to life? Relationalities in here and out there: A reply to Horton and Kraftl". *Area* 40: 289–292.

Hutta, Jan Simon and Carsten, Balser (Carla La Gata). 2013. "Identidades e cidadania em construção: historização do 'T' nas políticas de antiviolência LGTB no Brasil". In *Geografias malditas: corpos, sexualidades e espaços,* edited by Joseli Maria Silva, Marcio Jose Ornat and Alides Baptista Chimin Junior, 311–338. Ponta Grossa: Todapalavra.

IBGE. 2013. *Pesquisa nacional por amostra de domicílios.* Rio de Janeiro: IBGE.

Johnston, Lynda and Longhurst, Robyn. 2010. *Space, place and sex: Geographies of sexualities.* Lanham, MD: Rowman and Littlefield.

Lee, Debora (Jerri Adriano Comassetto Machado). 2013. "A geografia de uma travesti é uma barra, é matar um leão a cada dia". In *Geografias malditas: corpos, sexualidades e espaços,* edited by Joseli Maria Silva, Marcio Jose Ornat and Alides Baptista Chimin Junior, 27–38. Ponta Grossa: Todapalavra.

Longhurst, Robyn. 2001. *Bodies: Exploring fluid boundaries.* London: Routledge.

McCall, Leslie. 2005. "The complexity of intersectionality". *Signs: Journal of Women, Culture and Society* 30, n. 3: 1771–1800.

Namaste, Viviane K. 2000. *Invisible lives: The erasure of transsexual and transgendered people.* Chicago, IL: University of Chicago Press.

Nast, Heidi J. and Pile, Steve. 1998. *Places through the body.* New York: Routledge.

Nash, Catherine J. 2010. "Trans geographies, embodiment and experience". *Gender, Place and Culture* 17, n. 5: 579–595.

Nikaratty, Leandra. 2013. "O que mais me marcou na vida é ser barrada e não poder entrar nos lugares: esta é a geografia de uma travesti". In *Geografias malditas: corpos, sexualidades e espaços,* edited by Joseli Maria Silva, Marcio Jose Ornat and Alides Baptista Chimin Junior, 239–254. Ponta Grossa: Todapalavra.

Ornat, Marcio Jose. 2009. "Espacialidades travestis e a instituição dos territórios paradoxais". In *Geografias Subversivas: discursos sobre espaço, gênero e sexualidades,* edited by Joseli Maria Silva, 177–210. Ponta Grossa: Todapalavra.

Pain, Rachel. 1997. "'Old age' and ageism in urban research: The case of fear of crime". *International Journal of Urban & Regional Research* 21, n. 1: 117–128.

Pain Rachel. 2001. "Gender, race, age and fear in the city". *Urban Studies* 38: 899–913.

Pain, Rachel; Mowl, G. and Talbot, C. 2000. "Difference and negotiation of 'old age'". *Environment and Planning D: Society and Space* 18: 377–393.

Pelúcio, Larissa. 2009. *Abjeção e desejo: uma etnografia travesti sobre o modelo preventivo de aids*. São Paulo: Annablume.

Peres, Wiliam Siqueira. 2005. *Travestis Brasileiras: Construindo identidades cidadãs: movimentos sociais, educação e sexualidades*. Rio de Janeiro: Garamond Universitária.

Pile, Steve. 1996. *The body and the city: psychoanalysis, space and subjectivity*. New York: Routledge.

Ramirez-Valles, Jesus. 2016. *Queer aging. The gayby boomers and a new frontier for gerontology*. Oxford: Oxford University Press.

Riquelme, Fernanda. 2013. "A vida da travesti é glamour, mas também é violência em todo lugar". In *Geografias malditas: corpos, sexualidades e espaços*, edited by Joseli Maria Silva, Marcio Jose Ornat and Alides Baptista Chimin Junior, 55–68. Ponta Grossa: Todapalavra.

Rodó-Zárate, Maria. 2013. "Developing geographies of intersectionality with relief maps: Reflections from youth research in Manresa, Catalonia". *Gender, Place and Culture* 21, n. 8: 925–944.

Rodó-Zárate, Maria. 2015. "Young lesbians negotiating public space in Manresa: An intersectional approach through places". *Children's Geographies* 13, n. 4: 413–434.

Rose, Gillian. 1993. *Feminism & geography: The limits of geographical knowledge*. Minneapolis: University of Minnesota Press

Santos, Paulo Reis dos. 2010. "Desejos, conflitos e preconceitos na constituição de uma travesti no mundo da prostituição". *Revista Latino-americana de Geografia e Gênero* 1, n. 1: 39–48.

Silva, Edson Armando and Silva, Joseli Maria 2016. "Ofício, Engenho e Arte: Inspiração e Técnica na *Análise de Dados Qualitativos*". *Revista Latino-americana de Geografia e Gênero* 7, n. 1: 132–154.

Silva, Hélio. 1993. *Travesti: a invenção do feminino*. Rio de Janeiro: Relume-Dumará.

Silva, Hélio. 1996. *Certas cariocas: travestis e vida de rua no Rio de Janeiro*. Rio de Janeiro: Relume-Dumará.

Silva, Hélio. 2007. *Travestis: entre o espelho e a rua*. Rio de Janeiro: Rocco.

Silva, Joseli Maria. 2009. "A cidade dos corpos transgressores da heteronormatividade". In *Geografias Subversivas: discursos sobre espaço, gênero e sexualidades*, edited by Joseli Maria Silva, 135–150. Ponta Grossa: Todapalavra.

Silva, Joseli and Ornat, Marcio Jose. 2014. "Intersectionality and transnational mobility between Brazil and Spain in travesti prostitution networks". *Gender, Place and Culture* 22: 1073–1088.

Siverskog, Anna. 2014. "They just don't have a clue: Transgender aging and implications for social work". *Journal of Gerontological Social Work* 57, n. 2–4: 386–406.

Siverskog, Anna. 2015. "Ageing bodies that matter: Age, gender and embodiment in older transgender people's life stories". *NORA – Nordic Journal of Feminist and Gender Research* 23, n. 1: 4–19.

Valentine Gill. 2003. "Boundary crossings: Transitions from childhood to adulthood". *Children's Geographies* 1, n. 1: 37–52.

Waitt, Gordon and Gorman-Murray, Andrew. 2007. "Homemaking and mature age gay men 'down-under': Paradox, intimacy, subjectivities, spatialities, and scale". *Gender, Place and Culture* 14, n. 5: 569–584.

Williams, Mark E. and Freeman, Pat A. 2007. "Transgender health". *Journal of Gay & Lesbian Social Services* 18, n. 3–4: 93–108.

Witten, Tarynn M. 2009. "Graceful exits: Intersection of aging, transgender identities, and the family/community". *Journal of GLBT Family Studies* 5: 35–61.

Witten, Tarynn M. 2014. "It's not all darkness: Robustness, resilience, and successful transgender aging". *LGBT Health* 1, n. 1: 24–33.

Witten, Tarynn M. 2016. "Aging and transgender bisexuals: Exploring the intersection of age, bisexual sexual identity, and transgender identity". *Journal of Bisexuality* 16, n. 1: 58–80.

5 Social Inclusion

Measuring the Invisible and the Insignificant

Chloe Schwenke

There is no shortage of research ideas when it comes to evaluating the plight of sexual and gender minorities – lesbian, gay, bisexual, transgender, queer and questioning (LGBTQ+) persons and groups – around the developing world. The need for such research is clear and the stakes are high; members of sexual and gender minorities are generally excluded from many of the essential benefits of international development. They are frequently unable to secure decent employment, access health care and education, participate in democratic processes, find secure housing, get married or adopt children. They frequently face violence and abuse, and have limited if any access to justice or to the rule of law. Despite these many hardships, international development investments and programming largely excludes them as beneficiaries – for the simple reasons that they are underreported or invisible in the context of available data, or are intentionally ignored due to their marginalized status. The relegation of marginalized persons and groups to a low priority status for assistance is largely driven by political ideologies, and is beyond the scope of this chapter to address. Yet results always matter; under current operational norms among most of the prominent foreign assistance donors, the flow of funding for the design and implementation of international relief and development initiatives is tied to the ability of such aid programs to demonstrate results achieved. Donor funds are always scarce and competition to access them is therefore intense; only those projects that can demonstrate genuine progress compared to an empirically robust baseline stand a chance of being funded.

To a very large extent (and excluding data specific to the HIV epidemic), such baseline data are neither collected nor maintained with respect to members of sexual and gender minorities. In fact, in 65 countries sexual minorities are targeted by the law and in ten others, homosexuality is punishable by death; many LGBTQ+ people therefore exist in the shadows.[1] The collection of such data in these 75 countries is admittedly highly problematic.

Conceptually and morally, it isn't difficult to assert that there is a clear violation of fundamental human rights affecting persons who are excluded – by nature of characteristics which they have no agency over – from access to security, public services, economic and democratic participation, opportunities and freedoms enjoyed by other citizens. Many researchers approach the

DOI: 10.4324/9781315613703-5

measurement challenge with that intention, that is to demonstrate clear violations of human rights targeting LGBTQ+ persons and groups. In other
approaches, researchers seek to demonstrate that the violation or non-recog-
nition of these human rights generates adverse consequences on the economy
of the nation concerned, that is that exclusion penalizes not only those who
are targeted but also the entire economy. This type of research seeks to make
the case that exclusion of certain persons and groups is not in the common
economic interest of the majority (with no reference to any moral arguments)
and especially when there is a sufficient population of persons in these excluded
categories such that their exclusion can be shown to have a statistically significant negative impact on the economic indicators of the nations as a whole.[2]

Arguably, the nature of exclusion encountered by many persons who are
self-identified as LGBTQ+ extends significantly beyond denial or abuse of
human rights, as bad as that might be. The intentionality of exclusion of certain minorities, coupled with high rates of violence, castigation, humiliation,
stigmatization, reprobation, scorn and other forms of vilification directed at
LGBTQ+ persons, constitutes a profound assault on their human dignity, and
can be deeply corrosive of their individual sense of self-respect.

But are we able to measure an assault on human dignity? If so, will such
findings on dignity be more informative, persuasive or motivational than any
assessment of human rights abuses or infringements? Should we instead be
content to evaluate the quality of a truly human life on the degree to which
human rights (with no exceptions) are recognized and respected as entitlements
for all? What will that tell us? Or ought we to view human rights as existing
instrumentally so that their fulfillment makes a life of embodied human dignity
possible?

Before considering further whether the validation of universal human dignity ought to be the ultimate goal of social inclusion, it is helpful and arguably
more urgent to consider the current state of affairs in carrying out research on
and meeting the needs and aspirations of LGBTQ+ persons around the world.

Development as Measurable Progress

Arguably all rational human beings seek a truly human life,[3] in which their
essential needs are able to be met, where a secure and peaceful life of meaning
is possible, and where life is characterized by freedom, choices and opportunity.
Each individual and each society are likely to define the specific contours of
their own notion of a truly human life somewhat differently, however, which
complicates any effort to determine and compare whether and how much
progress is being achieved through international development assistance, local
development efforts or through changing social norms such as expanding levels
of social and economic inclusion. In the context of the development needs and
aspirations of LGBTQ+ persons, this situation begs the foundational questions
of how such "progress" is defined and by whom, what such progress is measured against, who the intended beneficiaries are and why, and what to do in

situations where same-sex sexual orientation or non-conforming gender identities are deeply stigmatized or even sanctioned as illegal and immoral.

A more difficult research environment is hard to imagine, but there is very little time in which to muse on theoretical possibilities. After a decade of civil society activism, advocacy and awareness raising across the globe on sexual orientation, gender identity and gender expression (SOGIE) issues, there is now a significant backlash evident in an increase in harassment, stigmatization, discrimination, exclusion, and violence against LGBTQ+ persons and groups (Encarnación 2019). If key global and national roleplayers are to be persuaded or cajoled into action to support sexual and gender minorities in the face of increasingly dire conditions, we need the essential data to support, justify, guide, measure and sustain appropriate interventions. And, we need it now.

While it is customary to treat sexual and gender minorities as a homogenous demographic, in reality the experiences and needs of each respective "letter" can vary widely. Those on the sexual orientation "SO" end of SOGIE, that is gay, bisexual or lesbian persons, are far from a monolithic group in terms of their daily realities and societal attitudes toward them. Gay men in particular can find themselves the target of intolerant and homophobic powers within faith-based communities, state and society. Ironically their relatively high levels of incidence of HIV/AIDS and other sexually transmitted diseases have made gay men[4] a focus of global programs aimed at stopping such threats to health. Anecdotal and some empirical findings indicate that such funding in turn has contributed to the ability of gay male communities and gay male civil society organizations to build their own capacity to organize, advocate, and connect through social media and strengthen their security, access to health care and other public services while also improving their leadership and livelihoods (Funders for LGBTQ Issues 2017; amfAR and The Foundation for AIDS Research and The Global Forum on MSM & HIV). While gay men remain highly vulnerable to public discrimination, violence and exclusion, they are now generally far better organized and with greater organizational capacity than at any time in the past.

For lesbians and bisexual persons and groups, the situation is often quite different. There is very little data available about bisexual persons, but strong empirical evidence[5] shows that lesbians have not been among those who are most vulnerable to HIV infection – generally referred to as a "key population" (UNDP Governance) – and are hence not part of the focus in the global fight against HIV. This has meant that compared to gay men, lesbians have not received equivalent funding support, and have not enjoyed commensurate access to the resources needed to build their own capacity as advocates and organizers. In nearly all cases, lesbians live in highly patriarchal societies anyway, so as women they are traditionally and systemically constrained from having their voices and priorities equally heard or acted upon even within the larger demographic of sexual minorities. In terms of security and safety, lesbians in some cultures enjoy a modicum of anonymity and space under the set-apart world of women. Still, even in such woman space, they are frequently very circumspect in being out about their orientation, but it would appear that the threshold for tolerance of intimacy between women is arguably higher than it is for men. Such generalizations however are problematic; incidences of honor killing of lesbians are not rare in many Islamic countries, and many societies outside the Islamic world also demonstrate significant intolerance for lesbians (Brownworth 2015; Martin et al. 2009; Mkhize et al. 2010; International Gay and Lesbian Human Rights Commission 2019).

There remains little civil society presence in any organized way for bisexual persons, although that situation is beginning to change. For now, bisexuals in the Global South are generally identified as either straight (i.e., they remain closeted) or as homosexual. The notion that one person can be intimately attracted to persons of the opposite, the same or a non-binary sexual identity is in its infancy in the Global South.

Turning to the gender identity and expression "GIE" end of the SOGIE issues collective, the prominent issue is not sexual orientation but gender identity. Intersex people constitute a very small minority and are frequently set apart from others on the LGBTQ+ spectrum due to the unique nature of their "non-conforming" gender identity. Some societies medicalize this condition, and while this doesn't obviate stigma and discrimination, it does sometimes mean that intersex persons are not seen by the public as "willful" deviants – as gender minority persons are often viewed.

Setting aside the complex phenomenon of gender expression, which in non-conforming situations (e.g., effeminate acting men, masculine acting women) can frequently lead to discrimination or worse, the remaining letter is the "T" for transgender. We do now have enough data, even if much of it anecdotal, to know that this demographic is the worst affected by discrimination, violence, stigma and exclusion of all within the SOGIE tent. In particular, the violence directed at transgender women is frequently extreme, and often deadly (HRC 2017). Yet, in the spirit of the canary in the coal mine (Doan 2001), that is the worst case situation, and in recognition that this author's own identity and experience is that of a transgender woman, the remainder of this chapter shall focus on the research needed to substantiate the transgender narrative associated with a truly human life of dignity in terms that can catalyze appropriate support, understanding and care. In other words, how do we measure and evaluate the plight of transgender persons, especially in the Global South?

The Research Challenge

While the data are neither definitive nor international, we can roughly extrapolate from US data that there are few transgender people in the world. It's a small population, estimated by the Williams Institute at a rate of 0.03% of the adult population within the United States, excluding those who identify as cross-dressers or gender non-binary (Flores et al. 2016).[6] With scarce funding for international development assistance and humanitarian relief and similarly scarce funding for basic research, it can be a very challenging argument to assert priority for such an insignificantly small population. That statistical insignificance alone condemns transgender people and their priorities to widespread exclusion and invisibility, unless a case can be made that the plight of this small population has importance beyond their numbers.

Does it matter to the economy if transgender persons are excluded, that is, that their human rights are ignored or abused and their human dignity not respected? The role of transgender people within an economic development agenda is almost entirely ignored. While some bilateral aid agencies like Swedish SIDA have incorporated LGBTQ+ communities as recipients of aid for many years, this has been the exception to the rule as opposed to the norm. And like most donors who do carry out work with sexual and gender minorities, there is little or no specific attention directed at transgender persons within the larger LGBTQ+ community. Similarly, the role of LGBTQ+ people as contributors to a nation's economic development has not been seriously examined, nor has their ability to access any of the benefits of economic development.

In recent years, a few more institutions such as the World Bank, various UN agencies and some bilateral aid agencies have become more engaged in support to research directed at sexual and gender minorities, but with very little specific systematic attention to transgender persons and groups. Consequently, while work on establishing theoretical frameworks and related empirical analysis is in its infancy across the whole LGBTQ+ spectrum, to date there is little or no recognition of gender identity issues.

The majority of economic impact research carried out to date has focused exclusively on the larger population of gay, lesbian and bisexual persons. Research carried out by the World Bank and the Williams Institute shows how discrimination and violence against LGB persons and groups have a detrimental impact on micro and macroeconomic development (Badgett et al. 2019). On a micro level, such exclusion has been shown to create an inequality of opportunity and thus detrimentally limit one's ability to access the benefits of economic development. This has two potential impacts on the individual. First, it creates a cycle of poverty fueled by violence, discrimination and a lower socioeconomic status. Second, it can limit the access to assets and resources that are essential to shared prosperity (Hawkins et al. 2014). Yet as important as this information is to address sexual minorities in general, the reality is that when the far smaller "n" of the transgender population and gender minorities is substituted, their impacts on the economy are likely to be marginal and insignificant.

Economies exist for people, and not the other way around. While it may be illuminating to know that the exclusion of lesbian, gay and bisexual persons incurs a negative consequence on the economy, the ultimate concern must be on how exclusion impacts the flourishing of all LGBTQ+ persons. To ascertain that information, particularly for the highly vulnerable and beleaguered transgender population, it is first necessary to find them. That is a daunting challenge given that very few countries legally recognize their existence; in one recent study, only 61 countries were found to have any provision to enable a transgender person to legally register their authentic name and gender marker.[7] Few countries include "transgender" in their census categories, and even the definition of what constitutes "transgender" is only now beginning to approach a consensus internationally. Some countries now accommodate a "third gender" or "other" gender category in official documents and census data, yet many transgender people do not seek such a status and only wish to be accepted in their claimed gender identity within the traditional gender binary – which they are not allowed to do. In short, transgender persons are frequently invisible.

There are reasons for this invisibility. First, the majority of research on LGBTQ+ populations is being carried out by gay and lesbian researchers, and their life experiences incline them toward an initial focus on issues of sexual orientation. In the context of HIV-focused research, until quite recently transgender people were not even recognized as a distinct sub-population, despite the fact that globally transgender women have now been shown to have the highest viral incidence rates of all (Baral et al. 2013; Jaspal et al. 2018; Singh and Krishan 2015; Williams et al. 2016). In addition, transgender persons are difficult (and expensive) for researchers to safely access, data collection agencies (including police departments and health institutions) generally do not bother to collect data specifically on transgender persons or crimes against them. Assumptions are made, if not always declared formally, that issues about gender identity will simply have to ride the coattails of research initiatives focused on sexual orientation. Such coattails belong to people heading in a different direction.

Another significant research challenge is that the largest research institutions and development organizations already have invested significant funds in gathering data that has been measured at the household level. Reaching LGBTQ+ persons in general, and transgender persons specifically, through household level data is problematic. Many LGBTQ+ persons are cast out of their homes, and even for those who remain there is widespread stigma attached to such status. Given this prevailing stigma and exclusion, the respondent in the typical head-of-household survey methodology is unlikely to report on LGBTQ+ family members, or to report on the authentic gender of any transgender members of that household. Few surveys have the financing or the motivation to reach transgender or other LGBTQ+ persons living on the street, at no fixed abode. Also, with such a large existing investment already made in the acquisition of household level data, there is a natural reluctance among such organizations and institutions to pursue expensive individual level data collection.

Other research constraints also apply. In many cultures, definitions of gender identity are based more on gender roles assigned at birth rather than on individual gender sensibilities regarding one's authentic identity. Not only must transgender persons in such societies struggle to have their individual gender identity recognized, they must also fight to find a culturally recognized gender role that will align with their own sense of integrity and authenticity.

Research requires finding subjects, and yet finding transgender persons in many societies who are willing to consent to participate in research projects is not easy. Many transgender persons are either deeply closeted, still living uncomfortably day to day in their birth-assigned gender while knowing that their authentic gender identity is otherwise, yet they fear the consequences of coming out. Other transgender persons are able to navigate their societies successfully (socially at least; their legal papers remain problematic) in their authentic gender identity without raising any suspicion that they once held a different gender marker. Such "stealth" transgender persons fear the consequences of owning their transgender status for any research project, lest the confidentiality of the study be breached.

Researchers do have techniques to work with hidden populations. Snowball sampling can leverage the contacts of a small group of accessible transgender persons, using their networks to reach larger populations. Such social networking incurs bias due to complex relationships, but techniques such as Douglas Heckathorn's respondent-driven sampling (RDS) are able to offer mathematical corrections to minimize the impact of such bias (Heckathorn 1997). Such techniques however are not inexpensive.

Added to these many challenges, transgender persons are generally aware of their status (even if frequently lacking the vocabulary to describe it well) at a very young age, often as young as 5. Were researchers to seek subjects who are transgender children, ethical practice demands that they first obtain parental consent. In societies that are suspicious, uncomfortable or deeply intolerant of children's claims to non-conforming gender identity, obtaining such consent will be very difficult.

Finally, in most of the world, the legal status of transgender persons remains unresolved, or is deemed illegal. Often, transgender persons are simply conflated with the gay community, even though they may be heterosexual or asexual in their orientation. That conflation may have dire consequences; as noted before, being gay or engaging in gay sexual behavior is illegal in 75 countries. Trying to reach research subjects who are classified as illegal is no minor hurdle, and keeping any data collected absolutely safe and anonymized is essential.

Currently there is a scarcity even of rudimentary baseline data pertaining to the lived realities of LGBTQ+ persons around the world, and while there is a growing chorus demanding funding to collect such data, such funding remains a low priority. In addition, such approaches are characterized more for their research limitations, questionable assumptions and missed sub-populations than for their robustness, and they largely leave out or make invisible the plight of transgender and non-binary persons.

Moral Concerns

As noted earlier, it is an arguable standard to structure any research carried out on the quality of life, prospects, freedoms and agency of any persons or groups on the basis of a human rights framework. Several such frameworks exist, spanning from the 30 articles of the Universal Declaration of Human Rights to Amartya Sen's distillation of the UDHR into seven freedoms (UNDP 2000).

While these delineations of claims and obligations have considerable moral weight and are reinforced by a robust system of international laws and treaties, the dynamic sensibility of being free to achieve a truly human life is not apparent on the face of it. For people who face lives constrained by exclusion, abuse and reprobation, freedom is a rare experience. Such excluded persons lack the capability to achieve truly human lives, and while a measurement based on the degree to which human rights are or are not satisfied would provide a snapshot of their quality of life, it would not answer important deeper questions of freedom, choice and capability.

To capture that set of sensibilities, it may be desirable to employ the capability approach as developed by such philosophers as Amartya Sen,[8] Martha Nussbaum, David Crocker, Sabina Alkire, and now refined and expanded by many others.[9] The capability approach has also been shown to be an effective framework in the measurement of the impacts of poverty and deprivation, although there are few if any examples of its use to evaluate social inclusion and the impact of inclusion and exclusion on human flourishing in terms of people's capabilities, that is, their actual opportunities to do and be what they have reason to value. Currently, many decision-makers within public policy use this approach in many contexts, from welfare economics, development policy, and in more academic exercises like social or political philosophy (Stanford Encyclopedia of Philosophy 2011). Recently, however, the capability approach has been posited as a reasonable framework upon which to base an LGBTQ+ (not necessarily including intersex or non-binary) economic development agenda (Park 2016).

Instead of focusing on traditional macroeconomic indicators of economic growth as the normative marker of "development", the capability approach focuses on the well-being, freedom and justice for individuals as well as groups. Development, in fact, is perceived as the freedoms of people, "development as freedom" (Sen 1999). In essence, the ultimate goal of development should be to cultivate "a process of expanding the real freedoms that people enjoy" (Sen 1999: 3).

Freedoms, capabilities and functionings form the core of the capability approach. The freedoms that people can enjoy are numerous, as well as diverse. Sen focuses on (at least) five main types when examining the "instrumental" perspective: political freedoms; economic facilities; social opportunities; transparency guarantees and protective security. All are highly relevant to LGBTQ+ persons and groups. When cultivated and given the opportunity to be experienced by such individuals and groups, these freedoms not only complement each other but also other serve to enhance overall human freedom in general (Sen 1999). In this sense, they are not only the end goals of development but also the very means by which to develop and experience freedom. Freedoms can be broken down further into two component parts: processes that enable free actions and decisions, and opportunities and choices that people have (as associated with their personal and social situations).

The expansion of a person's capabilities is also important, if people seek to lead the kind of lives they value. This would embrace the desire of transgender persons to live their lives within their authentic gender identity, while at the same time being accepted and "included" by society. Individual freedom is also crucial to development. According to Sen, freedom opens the door to the two-way relationship between the expansion of capabilities and public policy – driven by a progressive realization of the public's capabilities. In this sense, "development as freedom" has two component parts that are driven by freedom and that can be utilized by development professionals: *evaluation* and *effectiveness*. With the former component, a society should be evaluated by the freedoms and choices that its people enjoy – rather than on traditional indicators of macroeconomic growth. Such freedoms and choices will be heavily influenced by the degree to which such persons are included and their dignity is respected. With the latter component, freedom is also the main determinant of "individual initiative and social effectiveness" (Sen 1999: 19) although arguably such freedom is contingent on a cultural environment of social inclusion and acceptance.

The capability approach primarily is a way to frame development as a process of expanding the real freedoms and choices that people enjoy. In this approach, expansion of freedom is viewed as both (1) the *primary end* and (2) the *principle means* of development. The former is a "constitutive role" and the latter is an "instrumental role" regarding the role of freedom and choice within development.[10] Under the framing of the capability approach, human functionings consist of "beings and doings". A "being" is a state of being (well-nourished, educated, included), while a "doing" refers to an action (voting, participating in the market, working in a job). Capabilities are understood as opportunities to achieve important beings and doings. Capabilities refer to a person's actual freedoms or opportunities to achieve functionings (Stanford Encyclopedia of Philosophy 2011). For people who are systematically excluded, the application of the capability approach's comprehensive or holistic approach will make explicit which *sets* of capabilities are open to members of sexual and gender minorities or other excluded populations under study (Stanford Encyclopedia of Philosophy 2011).

Agency, Human Dignity and Human Rights

One very significant aspect of social inclusion is human agency – the recognition that human beings are equal in dignity, possess the capacity to reason and have a legitimate stake in influencing decisions in those aspects of their lives that will most directly affect them. Good governance is built on the premise of individuals as moral agents – rational and well-informed people with voice, who are afforded the opportunity to participate in deliberative processes in a way that reflects respect for their dignity, and that is responsive to their stake in the outcomes. People who are marginalized are generally excluded from such a participatory role, either by institutional processes that explicitly preclude them (e.g., by criminalizing or highly stigmatizing their status, as many countries do with respect to LGBTQ+ persons) or by their own internalization of pervasive societal messages of exclusion that diminish an LGBTQ+ person's own sense of worth and dignity, and effectively silence their voice in such deliberative spaces. Such self-censorship is often keenly felt by transgender persons, when their societies refuse to recognize their legal identity in their authentic gender, leaving them legally and economically as non-persons. Challenging the culture of social exclusion of sexual and gender minorities and particularly the highly vulnerable status of the transgender population will depend both on the creation of an enabling environment where diversity is accepted and valued, and on specific programmatic interventions to build that essential participatory capacity among the marginalized people affected. These outcomes will not be easy or quick to achieve, but for social inclusion to be embracing of all and significant in terms of measurable results, these objectives must be pursued.

The concept of human dignity is enigmatic. As leading theorist on the topic Arnd Pollmann noted, "There might already be an almost global consensus on the worth of the "dignity" concept as such, but by far no common *use* of the term" (Pollmann 2010: np). Still, there is an almost universal but intuitive sense of the significance of human dignity, which may be why it was chosen as the focal concept of the very first Article of the Universal Declaration of Human Rights: "All human beings are born free and equal in dignity and rights". In the context of SOGIE issues, this concept is particularly valuable, as captured by the political economist George Kateb: "The greater the suffering that a society may inflict on people within or outside its domestic jurisdiction, the more urgent the question of human dignity becomes" (Kateb 2011: 20). Already the existing anecdotal evidence is persuasive that the intensity of suffering inflicted upon transgender persons is egregious. For example, in countries such as Uganda, transgender women are frequently taken into custody by the police and then are forced to disrobe at the police station, in front of large crowds. Violence against transgender persons (especially against transgender women) is well documented in most countries where such records are kept. In the United States, transgender persons face enormous discrimination in their efforts to find and keep employment. As observed in the 2015 US Transgender Survey National:

> 30% of respondents who had a job reported being fired, denied a promotion, or experiencing some other form of mistreatment in the workplace due to their gender identity or expression, such as being verbally harassed or physically or sexually assaulted at work . . . 46% of respondents were verbally harassed and 9% were physically attacked because of being transgender . . . 10% of respondents were sexually assaulted, and nearly half (47%) were sexually assaulted at some point in their lifetime.
>
> (James et al. 2016: 1–11)

A vibrant debate continues as to whether human dignity is an innate and equal quality of all persons, or instead exists as a fragile potential for "embodied self-respect" – a potential that has to be fulfilled and self-actualized by the persons in question themselves under occasionally precarious life conditions. In the first view, human dignity serves as the foundation for all human rights.

In the latter view, human dignity can be interpreted as the sum of all human rights, once realized. In either sense, human dignity remains a valuable concept closely linked to human rights, and dignity provides a sensibility that can frame SOGIE issues in a dimension of social inclusion and human rights that is widely persuasive.

Human rights remain important. Internationally, human rights principles exert considerable influence (admittedly diminished under the previous American political administration in power) on international development both in terms of ethical theory and in the context of international laws, protocols and agreements. In both contexts – principles and laws – it is stressed that non-discrimination, equality and inclusiveness ought to underlie the practice of development, and that all people ought to be protected against egregious human rights violations. Views vary by country however regarding what constitutes a human rights violation, and the weight that should be accorded to the promotion of human rights values in the context of development.

Conclusions

The importance of the small transgender population rests not in their numbers, but in the opportunity which they present to the world to demonstrate a commitment to human dignity. The claims of transgender persons are fundamentally about inclusion – asserting that their humanity and dignity be recognized and respected, but only in the context of an identity that they feel is central to their access to freedom and meaning. Existing norms of exclusion and reprobation are perceived as a direct and egregious assault on their human dignity, and a deeper challenge to the premise of universality that is inherent to the concept of human dignity.

To address the needs and aspirations of excluded minorities such as transgender persons, the broader LGBTQ+ community and other vulnerable and marginalized groups (persons with disabilities, ethnic and racial minorities, indigenous populations), we must work together to make their plight explicit and comprehensible. This will require reliable and appropriate data, and the baselines and progress evaluations that flow from that starting point.

Data on the impact of social exclusion and on the many values inherent in social inclusion will only be meaningfully accessed with analytical frameworks that capture the internalized and lived realities of those who are most adversely affected by exclusion. To that end, the capability approach may offer a unique set of concepts, norms and principles upon which any effective, new social inclusion index might be devised. The insights and lessons that can be gleaned from the application of such an index would potentially have wide and far-reaching benefits to the design of policy and programs that contribute toward the goal of human flourishing for all – no exclusions. While progress has been made jointly at the World Bank and at the UNDP to generate such an index, it has yet to become operational in any significant way. Without research results based upon such a robust index, development funders (foundations, national donors, international donors) lack any way to establish a baseline of conditions that accurately describe the realities faced each day by transgender persons. Lacking such baseline data, no way exists to demonstrate results achieved as a consequence of donor funding – and hence accountability and attribution are impossible to substantiate. With no way to show results tied to development investments, development funding will remain limited to relatively small contributions to civil society capacity building for transgender groups (and LGBTQ+ groups more generally) or be limited solely to health (especially HIV) research into "key populations".

Only when development funds are provided at an appropriate scale will the critical needs for improved education, job creation, specialized training, democratic participation, legal reform, policy reform and so many other aspects of what Martha Nussbaum calls a "truly human" life become achievable for marginalized transgender persons around the world.

Notes

1 The data on the illegality of homosexuality do not directly address transgender and intersex persons, although in many instances they are simply conflated with "gay" persons and subject to the same sanctions regardless of their sexual orientation. See Cameron and Berkowitz (2016).

2 Initial studies carried out by the Williams Institute do demonstrate just such an adverse effect on the larger economy due to the exclusion of lesbian and gay persons; very limited data however make it impossible to draw similar robust conclusions about such a correlation with transgender and intersex persons. See Badgett and Crehan (2017: 45).

3 The philosopher Martha Nussbaum is one of the first and leading minds in the formulation of the capabilities approach. Under that ethical framework, she argues that a "truly human" life is characterized not only by happiness and by being rational but also by having valuable capabilities in order to pursue very human functionings. Her illustrative list of ten essential capabilities is often held up to be a roadmap to what constitutes a "truly human" life, and includes such attributes as living with and for others, having adequate access to health resources in order to live a full human lifespan, being in a position to plan one's own life rationally, participating in work opportunities that are dignified, engaging in social and cultural participation, interacting in a relationship with the natural environment, and more. For Nussbaum, a life that is truly human must demonstrate that these capabilities are being satisfied; none of the ten can be traded off or ignored. See Nussbaum (2001).

4 For the purposes of this chapter, the group who self-identifies as "men who have sex with men" (MSM) but who do not necessarily self-identify as gay men are being grouped together under the "gay men" category.

5 Data from the United States indicate that female-to-female sexual contact is an inefficient route of HIV transmission when compared to male/male or male/female sexual contact. The CDC reports that there are no confirmed cases of HIV from female-to-female transmission. See Deol and Heath-Toby (2009).

6 A 2014 report by the CDC indicated that in the United States, 1.8% of adult men self-identify as gay, 0.4% as bisexual, 1.5% of adult women self-identify as lesbian, and 0.9% as bisexual. See Volokh (2014). There are also significantly larger estimates of the transgender population in the United States, but the differences mostly relate to how "transgender" is defined, and specifically whether cross-dressers and gender binary persons are included as transgender. See Doan (2016).

7 Transgender Europe studied 126 countries and found only 61 that allowed transgender persons the legal right to change their names to align with their authentic gender identity. This research however excluded many countries in Africa and in the Middle East – countries which are largely hostile to LGBTQ+ persons and hence unlikely to offer this legal name change option to transgender persons. See Transgender Europe (2019).

8 Amartya Sen is both an economist and a philosopher. He won the 1998 Nobel Prize in Economic Sciences.

9 Feminist philosophers often regard the capability approach favorably, since one of their main complaints about mainstream moral and political philosophy has precisely been the relative invisibility of the fate of those people whose lives did not correspond to that of an able-bodied, non-dependent, caregiving-free individual who belongs to the dominant ethnic, racial and religious group, and who are therefore among the socially included. The capability approach incorporates issues of diversity by allowing for an evaluation of issues such as social inclusion that may be important for groups such as sexual minorities, but not necessarily so for others.

10 When comparing constitutive to instrumental roles, the capability approach asserts that freedom constitutes the primary ends of development. In this context, development research and analysis should consider the role of exclusion, which is essentially an "unfreedom". This also differs significantly from the "instrumental" role that links freedom to economic indicators of growth in that it matters fundamentally; freedom and social inclusion (understood as an indicator of respect for universal human dignity) is an end of development.

References

amfAR, The Foundation for AIDS Research & The Global Forum on MSM & HIV (undated) Lessons from the front lines: Effective community-led responses to HIV and AIDS among MSM and transgender populations. Available from: www.amfar.org/ uploadedFiles/_amfarorg/Articles/Around_The_World/GMT/2014/GMTMSM%20 050114v320.pdf [accessed 19th September 2019]

Badgett, M.V.L. & Crehan, P. (2017) Developing actionable research priorities for LGBTI inclusion. *Journal of Research in Gender Studies*, 7(1), 218–247. Available from: www. addletonacademicpublishers.com/search-in-jrgs/3091-developing-actionable-research-priorities-for-lgbti-inclusion [accessed 8th September 2019]

Badgett, M.V.L., Waaldijk, K. & Muelen Rodgers, Yana van der. (2019) The relationship between LGBT inclusion and economic development: Macro-level evidence. *World Development*, 120(8), 1–14. Available from: https://doi.org/10.1016/j.worlddev.2019.03.011 [accessed 3rd September 2019]

Baral, S., Poteat, T., Strömdahl, S., et al. (2013) Worldwide burden of HIV in transgender women: A systematic review and meta-analysis. *Lancet Infectious Diseases*, 13, 214–222

Brownworth, Victoria A. (2015) Job discrimination, corrective rape & honor killing of lesbians marked a troubling year: Erasing Lesbians in 2014. Available from: www.lotl. com/News/Job-Discrimination-Corrective-Rape-Honor-Killing-of-Lesbians-Marked-a-Troubling-Year-266/ [accessed 8th September 2019]

Cameron, D. & Berkowitz, B. (2016) The state of gay rights around the world. *The Washington Post*. June 14, 2016. Available from: www.washingtonpost.com/graphics/world/ gay-rights/ [accessed 19th September 2019]

Deol, A.K. & and Heath-Toby, A. (2009) HIV risk for lesbians, bisexuals & other women who have sex with women. Women's Institute at Gay Men's Health Crisis. June 2009. Available from: www.gmhc.org/files/editor/file/GMHC_lap_whitepaper_0609.pdf [accessed 19th September 2019]

Doan, P.L. (2001) Are the transgendered the mine shaft canaries of urban areas? *Progressive Planning: Special Issue on Gender and Violence*, 146. Available from: www.plannersnetwork. org/publications/2001_146/Doan.html [accessed 19th September 2019]

Doan, P.L. (2016) To count or not to count: Queering measurement and the transgender community. *WSQ: Women's Studies Quarterly*, 44(3), 89–110

Encarnación, O. (2019) The global backlash against gay rights: How homophobia became a political tool. May 2, 2017. *Foreign Affairs*. September/October 2019. Available from: www.foreignaffairs.com/articles/2017-05-02/global-backlash-against-gay-rights [accessed 25th September 2019]

Flores, A.R., Herman, J.L., Gates, G.J. & Brown, T.N.T. (2016) How many adults identify as transgender in the United States? The Williams Institute. June 2016. Available from: https://williamsinstitute.law.ucla.edu/wp-content/uploads/How-Many-Adults-Identify-as-Transgender-in-the-United-States.pdf [accessed 25th September 2019]

Funders for LGBTQ Issues & The Global Philanthropy Project (GPP) (2017) 2015/2016 Global resources report: Government and philanthropic support for lesbian, gay, bisexual, transgender and intersex communities. Available from: https://lgbtfunders.org/wp-content/uploads/2018/04/2015-2016_Global_Resources_Report.pdf [accessed 16th August 2019]

Hawkins, K., Wood, S., Charles, T., He, X., Li, Z., Lim, A., Mountian, I. & Sharma, J. (2014) Sexuality and poverty synthesis report, IDS Evidence Report 53, Brighton, UK: Institute of Development Studies. Available from: http://opendocs.ids.ac.uk/opendocs/bitstream/handle/123456789/3525/ER53.pdf?sequence=1 [accessed 18th August 2019]

Heckathorn, D. (1997) Respondent-driven sampling: A new approach to the study of hidden populations. *Social Problems*, 44(2), 1 May 1997, 174–199. Available from: https://doi.org/10.2307/3096941 [accessed 25th September 2019]

Human Rights Campaign (2017) Violence against the transgender community in 2017. Available from: www.hrc.org/resources/violence-against-the-transgender-community-in-2017 [accessed 18th August 2019]

International Gay and Lesbian Human Rights Commission (now OutRight Action International) (2019) 15 targeted killings of lesbians in Thailand since 2006: Iglhrc report. Available from: https://outrightinternational.org/content/15-targeted-killings-lesbians-thailand-2006-iglhrc-report [accessed 18th September 2019]

James, S., Herman, J., Rankin, S., Keisling, M., Mottet, L. & Anafi, M. (2016) *The Report of the 2015 U.S. Transgender Survey*. Washington, DC: National Center for Transgender Equality.

Jaspal, R., Kennedy, L. & Tariq, S. (2018) Human immunodeficiency virus and trans women: A literature review. *Transgender Health*, 3(1), 239–250. Available from: www.ncbi.nlm.nih.gov/pmc/articles/PMC6323592/#B17 [accessed 26th September 2019]

Kateb, G. (2011) *Human Dignity*. Harvard: Harvard University Press.

Martin, A., Kelly, A., Turquet, L. & Ross, S. (2009) Hate crimes: The rise of 'corrective' rape in South Africa. *Action Aid*. Available from: www.actionaid.org.uk/sites/default/files/publications/hate_crimes_the_rise_of_corrective_rape_in_south_africa_september_2009.pdf [accessed 19th September 2019]

Mkhize, N., Bennett, J., Reddy, V. & Moletsana, R. (2010) *The Country We Want to Live in: Hate Crimes and Homophobia in the Lives of Black Lesbian South Africans*. Cape Town: HSRC Press.

Nussbaum, M. (2001) *Women and Human Development: The Capabilities Approach*. 1st ed. Cambridge: Cambridge University Press

Park, A. (2016) *A Development Agenda for Sexual and Gender Minorities*. Los Angeles, CA: The Williams Institute, UCLA School of Law. Available from: https://williamsinstitute.law. ucla.edu/wp-content/uploads/Development-Agenda-for-Sexual-and-Gender-Minorities.pdf [accessed 3rd September 2019]

Pollmann, A. (2010) Embodied self-respect and the fragility of human dignity: A human rights approach, chapter in *Humiliation, Degradation, Dehumanization: Human Dignity Violated* (pp. 243–261). Available from: www.researchgate.net/publication/321599404_ Humiliation_Degradation_Dehumanization_Human_Dignity_Violated [accessed 14th January 2019]

Sen, A. (1999) *Development as Freedom*. 1st ed. Knopf.

Singh, S., Krishan, A., et al. (2015) *Experienced Discrimination and its Relationship with Life Chances and Socio-economic Status of Sexual Minorities in India*. New Delhi: Amaltas Research; The World Bank Group. (Unpublished at time of writing, pending a World Bank peer review)

Stanford Encyclopedia of Philosophy (2011) *The Capability Approach*. Stanford, CA: Stanford University. Available from: https://plato.stanford.edu/entries/capability-approach/ [accessed 26th September 2019]

Transgender Europe (2019) Legal gender recognition: Change of name. Available from: https://transrespect.org/en/map/legal-gender-recognition-change-of-name/ [accessed 25th September 2019]

United Nations Development Program (UNDP) (2000) Human development report 2000, pages 3 and 146. Available from: http://hdr.undp.org/sites/default/files/reports/261/ hdr_2000_en.pdf [accessed 25th September 2019]

Volokh, E. (2014) What percentage of the U.S. population is gay, lesbian or bisexual? *The Washington Post*. July 15, 2014. Available from: www.washingtonpost.com/news/volokh-conspiracy/wp/2014/07/15/what-percentage-of-the-u-s-population-is-gay-lesbian-or-bisexual/?utm_term=.d3254e724b1f [accessed 14th September 2019]

Williams, H., Varney, J., Taylor, J., et al. (2016) *The Lesbian, Gay, Bisexual and Trans Public Health Outcomes Framework Companion Document*. Public Health England, London, United Kingdom, 2016. Available from: http://lgbt.foundation/assets/_files/documents/jul_16/ FENT__1469789610_PHOF_LGB&T_Companion_2016_FINA.pdf [accessed 26th September 2019]

6   Political Entanglements

Practicing, Designing and
Implementing Critical Trans
Politics Through Social Scientific
Research

Rae Rosenberg

Scholarly conversations and writing around queer methodologies have led to innovative interrogations and explorations of what social scientific research can look like with lesbian, gay, bisexual, transgender, queer, Two-Spirit and additional (LGBTQ2+) communities. Queer methodologies interrogate the ways in which identities become (re)constituted amongst researchers and participants, and in the process problematize subject positions and forms of power as they manifest throughout the research process. Yet few discussions have emerged around developing methodologies specific to working with trans communities, perspectives, politics and researchers, and the ways in which trans experiences, subjectivities and politics inform work with members of our own community.

This chapter discusses my experiences of conducting research with incarcerated trans feminine[1] individuals in the United States (US) in a project designed collaboratively with the Prisoner Correspondence Project (PrisCoPro), a Canadian activist collective that offers resources for incarcerated LGBTQ2+ people, and facilitates pen-pal correspondence between them and non-incarcerated LGBTQ2+ people in the US and Canada. The collective is run by volunteer "outside members" who spend an average of a few hours per week sorting through and responding to mail, cultivating resources and hosting outreach events to find new outside pen-pals. The PrisCoPro has a few hundred inside members, the majority of whom are incarcerated in the US. At the time of this research there was one PrisCoPro based in Montreal; since then a second branch was established in Toronto, however the two collectives operate independently of each other. Collective membership is open to all people, however almost all collective members identify as LGBTQ2+.

The research discussed in this chapter commenced in early 2012 and was based at McGill University. An invitation to participate in a long-answer, hand-written questionnaire was mailed to 29 trans feminine individuals in men's correctional facilities[2] in Texas, New York, Pennsylvania, California, Florida, New Mexico, Georgia and Indiana. These individuals were all inside[3] members of the PrisCoPro had self-identified as trans and/or non-binary and consented to receiving LGBTQ2+ content in the mail. Participants self-identified along a range of racial and ethnic identities, including White, Native American, Black/African

DOI: 10.4324/9781315613703-6

American, Latina/Hispanic and bi-racial (Rosenberg and Oswin 2015). In addition to PrisCoPro members, two participants mailed copies of the questionnaire to their friends and partners, while others sent me the names and addresses of individuals they knew who wanted to participate. Between the initial mail-out and this snowball effect, a total of 23 people participated in the research.

The following pages explore the ways in which a critical trans politics was built into my research methodologies and how this informed the design and implementation of the research. I illustrate how my positionality as a trans person informed the research process, from designing my methods to negotiating power differentials that emerged during the research. This is followed by an exploration of the bureaucratic challenges of obtaining ethical approval for my research, particularly given the conflicting goals between a trans and LGBQ2+ collective, university and state ethics boards. Lastly I explore the emotional labor involved in conducting research with multiply-marginalized trans people as a trans person, and the particular ways in which trans researchers bear witness and attend to systemic transphobia in the research process. Attending to these complex affective entanglements enables an engagement with forms of community accountability and healing, and embeds a multifaceted politics of care that challenges the necropolitical forces of transphobic state violence and intellectual voyeurism that can be produced in trans-focused research.

Trans Incarceration and Queer Research Methods

In order to gather data that reflects the intricacies of gender and power in carceral space, I utilized long-answer questionnaires to gather detailed, in-depth and subjective accounts of trans feminine prisoners' experiences of incarceration. The exact number of trans and/or non-binary people incarcerated in the US has been difficult to quantify due to the lack of statistical information and recognition of trans and/or non-binary identities within the Prison Industrial Complex (PIC). Scholars and activists, however, have identified an overrepresentation of trans people in the PIC due to the multiple forces that lead to higher risks of incarceration, particularly for trans women of color (Hagner 2010; Maruri 2011; Peek 2004; Spade 2011; Stanley and Smith 2011; Sylvia Rivera Law Project 2007). Trans people are often pushed into informal and criminalized economies due to a combination of transphobia among peers and within families, legal barriers for name and gender marker changes, and a prevalence and ambivalence toward transphobic employment discrimination.

I followed the guidance of feminist methodologies to maintain an awareness of my positionality and exercise reflexivity as a researcher, in order to prioritize

> non-hierarchical interactions, understanding, and mutual learning, where close attention is paid to how the research questions and methods of data collection may be embedded in unequal power relations between the researcher and research participants.
>
> (Sultana 2007: 375–376)

The practice and exercise of reflexivity informs a sense of mutuality and openness between researchers and their participants, and "looks both 'inward' to the identity of the researcher, and 'outward' to her research and what is described as 'the wider world'" (Rose 1997: 309), aiming to interrupt power relations throughout the research process. For my work, practicing reflexivity meant recognizing that although I am a trans person, I do not experience patriarchy, homophobia, racism and transphobia in the same way as do my research participants. It further encouraged me to consider my privileges as a researcher and as someone who was not, and never has been, incarcerated. Looking at these differences allowed me to explore the relationship between myself and my research participants, as well as how to form an accessible research method that would offer participants entry into the production of research-based knowledge. Maintaining this awareness was a necessary part of informing how my research was designed and implemented, how I would conduct my research analysis and maintain active allyship to incarcerated trans feminine persons, and how I would practice a critical trans politics through my research.

My research also drew on queer methodological research approaches. In its anti-normative position, Browne and Nash argue that "keeping queer permanently unclear, unstable and 'unfit' to represent any particular sexual identity is the key to maintaining a non-normative queer position" in research (2010: 7–8). This fuzziness posits a dilemma for the research process, as queer research cannot ever be a definable concept, yet is being utilized as a framework that centralizes "queering" as a research method. Browne and Nash (2010) note that it can be difficult to identify a queer method, because its unbounded nature allows it to be many things at once: it can problematize narratives of coherence and practices in social scientific research, while simultaneously essentializing what queering methods and queer epistemologies may look like, and thus become antithetical to queerness. Queer methods may not even exist at all, in the sense that all methods can be queered (Browne and Nash 2010: 12). For Gorman-Murray, Johnston and Waitt "a queer methodology must facilitate telling and interpreting narratives that do not inadvertently impose meanings rather than seeking to rework and create new narratives" (2010: 101). Queer methods must consequently remain open-ended, always positioned to destabilize and question practices that have come to be valued for certain knowledge production.

Additionally, queer research has been further considered as a political orientation that instructs research designs, aims and outputs of knowledge production. As Nash explains, the distinction of queer methodology is made "not only [in] its underlying theoretical, epistemological and ontological starting points but its political commitment to promote radical social and political change that undermines oppression and marginalization" (2010: 131). A queer methodology, then, is always charged with a particular politics that is rooted in the theoretical and historical meanings of queer, always turned toward new openings, playing with the production of knowledge and meaning, interrogating how identity is always being (re)constituted for researchers and participants, and shifting how subject positions and forms of power are understood as manifesting throughout the research process (Gorman-Murray et al. 2010; Taylor 2010).

Queerness in research practices has also been considered in the formation and utilization of queer ethics as a methodological approach (Detamore 2010). Detamore meditates on what the strategy of queerness can reveal about our methodologies by using queer ethics as a methodological framework to centralize a politics of intimacy that induces specific forms of knowledge production, relationships and experiences, explaining this as the "radical notion of a queer attachment to the bonds created through research" (2010: 179). These bonds emerge through the intimate attachments and affects that are formed between researchers and participants, and these intimacies create particular forms of queer knowledge. Queerness as an ethical political position is fundamentally tied to retaining critical trans and queer political orientations in the face of bureaucracy, and offers a reimagining of scholar-activist participation in knowledge production.

A queer ethics-as-method approach was particularly relevant to this research project, as one of the larger challenges involved navigating the bureaucracies of university and prison ethics boards. Prison institutions are intensely regulated and position researchers in highly securitized networks of guidance and surveillance that make it difficult to conduct research that prioritizes the safety and well-being of participants. For example, research involving trans women who were incarcerated in the California Department of Corrections and Rehabilitation (CDCR) found that the CDCR "does not employ an agreed-upon definition of transgender to identify or classify inmates" (Sexton et al. 2010: 840). In order to maneuver around this problem, researchers agreed to have CDCR correctional staff identify and recruit prisoners who they believed were trans. While this strategy allowed researchers to access more potential participants than individually contacting prisoners, accessing participants in this way was problematic for several reasons. In this process, many of those who had been identified as trans were queer cis men who did not identify as women or as non-binary. Misgendering gay men is not only an effect of homophobia and transphobia, but could risk these prisoners' safety by inciting transphobic reactions from other inmates. This misgendering process also revealed the tremendous ignorance within carceral institutions about trans and/or non-binary identities and the transphobic idea that trans women are not women.

In addition to the problematics of misgendering prisoners, having prison administrators identify who they thought were trans women meant that certain incarcerated trans women could have been omitted from this research study. The authors did not feel that this potential omission presented a systemic bias and, to the contrary, argued that CDCR administrators were over-inclusive in the selection process (Sexton et al. 2010: 843). However, given the disproportionate transphobic biases and violence at the hands of correctional officers, there is a strong possibility that certain trans women in this case were not allowed to participate in the research if they had a negative relationship with prison administrators (Spade 2011; Sylvia Rivera Law Project 2007).

Given the negative impacts that could so easily arise in research opportunities for incarcerated trans people, I was hyper-aware that the incarcerated trans people I was contacting may be wary of working with a researcher and have little trust in my ability to prioritize their safety and confidentiality. As someone who had never been incarcerated, I informed myself as best as I could of the issues of overwhelming harassment, violence, isolation, intimidation and coercion that incarcerated trans feminine individuals experience in order to design my research with care, sensitivity and safety (Hearts on a Wire 2011; Stanley and Smith 2011; Sylvia Rivera Law Project 2007). Partnering with the PrisCoPro helped to minimize the risk of assuming peoples' identities, outing a person without their consent, misidentifying someone as gay and/or trans or contacting an incarcerated person about a sensitive or undesired topic without their consent. Due to my involvement as a collective member prior to beginning the research, the larger collective felt comfortable allowing me access to their organizational database in order to cultivate a list of potential research participants. Of the PrisCoPro's few hundred inside members, almost all identified along the LGBTQ2+ spectrum and were incarcerated in Canada and the US. Approximately 50 of these members were identified as trans feminine, all of whom were incarcerated in the US. While I had initially hoped to work with a broader spectrum of trans and non-binary people, there were no incarcerated PrisCoPro members who identified as trans masculine or assigned female at birth; consequently, the focus of my research was limited to trans feminine individuals.

Incorporating Critical Trans Politics into Research Design

The PrisCoPro's online database of inside members was immensely helpful in aiding me to identify potential research participants. There were, however, still opportunities where I could potentially misidentify or miss a person, or accidentally mail someone who did not want to be contacted. To mitigate the chances of this occurring, two pieces of criteria were established to identify potential research participants: 1) the person would need to have indicated that they identified as trans and/or non-binary on their membership form and 2) the person would have had to consent to receive queer-related mailings from the PrisCoPro on their membership form. These criteria helped indicate that a person was "out", or at least comfortable receiving LGBTQ2+ content in the mail, and also lessened the chance of misidentifying someone.

Finding self-identified trans and/or non-binary members in the PrisCoPro's database was a nuanced process that involved complicated searches, both in the online and hard copy databases. Some individuals were easily identifiable if they had signed up using a chosen name that was distinctly different from their legal name (i.e., Sarah versus Jason). This was not completely reliable, however, because some trans members had already changed their legal names and thus did not have a different name displayed in the chosen name section of the database. As well, not all trans members had a different chosen name, and there were also cis members who had different chosen names. This also could not account for any human error made by PrisCoPro collective members when entering a chosen versus legal name into the database. In order to approach these complications, I employed a more in-depth database search of handwritten pen-pal biographies, emails with outside pen pals and written exchanges between inside members and the PrisCoPro collective to ensure that a person had self-identified as trans and/or non-binary. All collective members have access to these materials, as letters with inside members are kept within specific files to keep track of what resources they have been sent, any information regarding their pen-pal correspondence and any writing or artistic submissions for collective publications such as newsletters or zines. With these combined criteria I was able to identify 29 inside members who were trans and consented to receiving LGBTQ2+-related mail from the PrisCoPro.

While it would have been ideal to conduct in-person or phone interviews with participants, doing so is very costly, time-consuming, monitored and heavily restrictive in institutions that allow researchers to talk to prisoners. While it would have been more feasible to conduct interviews over the phone, they were not preferable because they are limited in time (approximately 15 minutes per phone call) and are also heavily monitored by prison administration and often recorded. Engaging in phone interviews could risk participants' comfort and safety, as well as undermine the data because participants might not feel comfortable sharing personal information and experiences over the phone as they might in a hand-written questionnaire.

Researchers have identified that mailing prisoners can be a more viable option for conducting work in a limited timeframe and/or large geographic area (Moran et al. 2009). For my work, sending research documents in the mail mitigated several of the aforementioned ethical concerns in conducting my research, notably identifying trans people and contacting them without their consent. Using questionnaires decreased the possibility of correctional staff intervening in the interview process or potentially retaliating against individuals for participating in the research, as I believed that the questionnaires were less likely to be read in detail than correctional staff listening to a verbal exchange. This still did not guarantee that incoming and outgoing mail would not be read by prison staff, which could risk participants' safety as well as impact what information participants were willing to share. I was surprised to find that this risk did not seem to censor participants, as most of them spoke freely about their experiences. Of the few who felt that they could not provide certain detail, they stated this plainly in their responses and even sent the information they excluded in separate letters, because they felt their letters would be read with less scrutiny than a formal-looking questionnaire. An additional benefit to using the questionnaire was that it could provide participants with more time to think about the questions and leave room to engage in self-care when answering difficult questions (Meth 2003).

All 29 potential participants were sent a mailing in June of 2012, which included a description of myself and the research project, an invitation to participate, two copies of the consent form, an in-depth questionnaire, ten stamps, three pieces of extra blank paper, three letter-sized envelopes and a manila envelope. Extra stamps and envelopes were provided in the event that participants would like to contact the McGill Research Ethics Board (REB), my supervisor or myself. If the invited participants chose to partake in my research, they were directed to keep one consent form and mail me the second, along with their questionnaire. Because I did not have access to a private mailbox at McGill University, participants were instructed to send their research materials to the PrisCoPro's office. From there, I transferred these materials to my locked office at McGill University where I stored them in a locked filing cabinet. Once the questionnaire was received, participants were compensated with $25, which was sent through a money transfer into their prison bank accounts. If they could not receive money transfers, individuals sent me information about purchasing items through commissary, or a store within correctional facilities, which was purchased for up to $25.

The questionnaire comprised a mix of 50 open- and closed-ended questions. These questions asked about various aspects of participants' lives in prison. Many of the questions, as well as the organization of the questionnaire, were influenced by Hearts on a Wire's (2011) publication of a similar survey with incarcerated trans feminine persons located in Pennsylvania prisons.[4] Because of my positionality as an educated, class-privileged, non-incarcerated person, I was uninformed about what kinds of questions to ask, how to ask them and how to present my research in a way that was accessible. The Hearts on a Wire (2011) questionnaire was a useful guide because it was written and designed by currently and formerly incarcerated trans people. As such, the language, formatting and focus of their questions were much more accessible and relevant than the questions and questionnaire design that I had initially drafted. With permission from Hearts on a Wire, I adapted many of their questions for the questionnaire that I developed for my research. Questions were grouped in themes and presented in the following order: gender, housing, gender expression, transitioning, harassment and violence, locations in correctional facilities, relationships and community, and struggle and resilience. Some of these sections proved to be more conclusive than others; in particular, "locations in correctional facilities" did not gather the kind of information I was seeking, most likely because the questions appeared to be unimportant and/or confusing to research participants.

Utilizing questionnaires as my main methodological approach presented its own forms of sample bias and limitations. Because the individuals who were invited to participate were involved in the PrisCoPro, my research could be excluding individuals with varying levels of literacy, competency with English and/or difficulty writing (Meth 2003). As well, contacting members of an organization biased my research population toward those who were already in need of support and could thus be more inclined to participate in a questionnaire, whereas conducting in-person interviews could access other trans and/ or non-binary people who perhaps experienced less mistreatment and were less inclined to reach out for support. This particular drawback regarding a selection bias "raises questions about validity and the 'truthfulness'" (Meth 2003: 202) in the use of in-depth writing as a research tool, as Meth noted in her account of using diaries as a qualitative research method. Meth (2003) also notes an additional problem with in-depth writing in its hyper-focus; particularly, that without the inability to raise additional questions through a conversation in a semi-structured interview, the data can be more limited and data collection less rigorous.

These limitations are valid concerns of using in-depth questionnaires as a methodological tool. That being said, contacting participants through the mail proved to be the best option due to the limitations and ethical concerns in conducting research with a highly vulnerable population like incarcerated trans feminine persons. While I did not ask participants why they chose to answer my questionnaire, many of them expressed that they were excited for an opportunity to have their voices heard and felt empowered by knowing that a researcher was interested in hearing about their experiences. Using emotional, highly personal writing as a form of qualitative research "offer[s] the opportunity for respondents to define the boundaries of their shared knowledge" (Meth 2003, 196), encouraging further engagement by participants through foregrounding their voices and involving them in the cultivation of knowledge production.

Ethical Dilemmas and Conflicts

Many concerns about involving incarcerated trans feminine people in research were held by McGill University's REB, which identified participants as a highly vulnerable population. Amongst concerns of maintaining the safety, confidentiality and well-being of my participants, a larger question emerged of how to stand by my participants if any issues were to arise in the research process, how my research and role as a researcher could impact the relationship between the PrisCoPro and its members, and how my own emotional well-being could be impacted by embarking on a research project involving stories of trauma and violence.

My primary concern surrounding participants' risks was the threat of retaliation by correctional staff or other inmates if the mail were to be intercepted and read, particularly if there was information that they did not want to be shared. While letter interception and the reactions of correctional staff or other inmates were beyond my control, I built some precautionary measures into the questionnaire design to mitigate the potential consequences of retaliation against research participants. Participants were able to select if they wanted to remain anonymous in any use of their questionnaires, and if they would prefer the use of a pseudonym in lieu of their name. Maintaining confidentiality in this way would help sustain not only a level of protection for participants in the report but also the relationship between the PrisCoPro and its inside members, which is based on solidarity and trust. The questionnaire began with guidelines for participants, which stated four reminders that would serve to reinforce participants' choice in what they wanted to write down, to remind them that they could take breaks if feeling stress or emotional harm from answering the questions, and also to reinforce that participants should not answer questions if they felt it could compromise their safety. The section of the questionnaire on harassment and violence had additional reminders, similar to the Hearts on a Wire (2011) survey, that the questions might be hard to answer and that participants should leave them blank if they felt that there might be a chance of retaliation if their mail was intercepted and read.

Dilemmas in Bureaucratic Gatekeeping

Receiving ethics approval from McGill University's REB was an arduous and lengthy process, taking a total of four months, from February to June of 2012. After months of a back-and-forth exchange about concerns regarding the protection of my participants, the largest issue was whether or not I needed to obtain permission to conduct my research from the research board of each separate correctional institution. I expressed that I would not seek approval from each institution's research board because doing so could significantly compromise participants' safety, and that my research would most likely not pass their research boards because of the anti-PIC politics of both the PrisCoPro and my own research proposal. As well, every state has its own requirements for researchers, including detailed paperwork, background checks and meetings with prison research boards, all of which could take between six months to a year before being granted approval. This was not viable due to the time constraints of a Master's degree. Following back-and-forth communication between myself and the chair of McGill University's REB about this dilemma, I received word that "[s]erious consideration must be given when weighing [my] arguments to override the research ethics safeguard procedures put in place by the prison authorities" (McGill University REB, personal correspondence, May 13, 2012), and that McGill University's REB planned to seek legal consultation about my request to bypass prison research boards. In response to this news, I cultivated the following formal response explaining my reasoning to bypass prison research ethics procedures, and sent these explanations to the REB.

1) *The potential rejection of my research due to stereotyping and bias against transgender people.* As research has identified, prison administrations often harbor transphobic sentiments. Because required background checks by correctional institutions would indicate that I am trans, I could be interpreted as inherently biased by their research boards and my research might be rejected. Furthermore, it might be assumed that my participants share my personal and political ideas, which could potentially create unsafe circumstances for them.

2) *The threat that obtaining approval by prison research boards might impose on the safety of my participants.* I was very concerned that submitting my research application and materials to correctional departments might create harmful

situations for my participants, as well as counter the measures I was taking to ensure their safety. If certain prison staff or administrators who had discriminated against trans prisoners became aware of my research, participants might experience increased mistreatment and harassment as intimidation to not disclose these negative experiences. Additionally, I was concerned that a general awareness of my research inside correctional facilities might further endanger participants from other inmates who became aware that they were receiving monetary compensation for participating.

3) *The probable rejection of my research due to correctional facilities' narrow interpretation of "beneficial research".* Almost every correctional research board stated that they approve research that contributes to the advancement of their institutions. I strongly believed that my research would be unfairly rejected for not benefiting their institution and the PIC more broadly. As the focus of my research would most likely take the form of critiquing correctional administrations' treatment of incarcerated trans people, my analysis and conclusion might not be viewed favorably by the administration. Furthermore, because incarcerated trans populations have reported frequent and extreme experiences of discrimination by correctional staff, the administration may not want my research to be conducted and reject my application as counter to the goals and desires of the institution.

4) *The possible negative impacts on the PrisCoPro.* Because the activist organization's name was embedded in my research, they themselves might be flagged with my research application and, consequently, their correspondence with certain prisoners could be disabled. This would be devastating for both the inside members who rely on the PrisCoPro to maintain their connection to the LGBTQ2+ community and queer networks of support, and the PrisCoPro which prides itself on maintaining consensual, respectful and safe communication with its inside members. As such, myself and the PrisCoPro agreed that I would not be allowed to contact their inside members if I had to receive approval from correctional research boards, and I would not be able to conduct my research as it was currently proposed.

After considering these arguments and obtaining legal consultation regarding my research, I received REB approval under the understanding that "there is a heightened and absolute obligation on (my) part to be aware of and to **immediately report any adverse events which may arise as a result of the study procedure**" (McGill University REB, personal communication, 12 June 2012, bold in original). To my knowledge there were no adverse events that occurred as a result of my research; alternatively, I received many positive responses both in my questionnaires and separately in the forms of thank you letters and friendship cards. As mentioned earlier, almost all participants seemed eager to share their experiences with me and appeared grateful that a researcher showed interest in their lives and well-being.

In exploring the bureaucracies of ethics boards, Detamore asks, "What happens when the ethical standards of bureaucracy do not fall in line with the ethical formations, norms and values of the human subjects we study?" (2010: 174). I would add to that question, what happens when these ethical standards do not align with our own ethical formations as researchers, and the ethical formations of other organizations we have aligned ourselves with? Do we relinquish our dedication to some aspects of our work to find common ground, simply to receive ethics approval? Or do we insist on our ethical position and refuse to compromise, standing ground in our political convictions and hoping that our persistence will eventually lead to ethics approval? In my experience I would suggest the latter – that in refusing to compromise we not only uphold our own ethical position as researchers based on a critical trans politics but that we also insist that the ethical standards of bureaucracy stretch the limitations of institutions to meet community-based research where it is situated. Otherwise, not only would our work inherently be compromised but also our words would be rendered meaningless and our positions as researchers antithetical to the meaning of queer.

Navigating Feeling

Initially I had made few considerations about my well-being in this research other than the potential of my name being flagged by prison administration and somehow getting passed over to law enforcement. However, throughout the research process, it became clear that the impact on my emotional health was much greater than I had originally anticipated. Since participants had hand-written their questionnaires, I had to read them and type them up before beginning the coding process. Through the acts of reading participants' writing, re-writing them into word documents, coding their experiences, analyzing their statements and re-reading their words repetitively throughout my writing process, a tremendous attachment was fostered between myself and them, and the affect of their writing was transmitted into me and became enmeshed in my emotional and psychological experiences.

Participants' tales of solitary confinement, having visitation rights revoked and property stolen, and experiencing verbal and violent harassment and sexual assault became ever-present, nagging thoughts. As Rager (2005) states, "qualitative inquiry is not a purely intellectual exercise but rather one for which researchers enter the world of their participants and, at least for a time, see life through their eyes" (24). While I had expected this kind of content in their writing, nothing had prepared me for the ways I would internalize their experiences, particularly regarding violence and sexual assault. This internalization became molasses, slowing me down as my body became a conduit of emotion from the questionnaires, becoming what Probyn describes as "the battleground where ideas and experiences collide" (2011: 89). Reading through such thoroughly documented experiences of self-harm, mental illness and transphobic trauma and violence had powerful impacts on me, inciting my own depression, anxiety and paranoia. Even though my research participants and their assailants were contained in concrete shells across the US, I felt so closely linked to my participants that their stories terrified me of my own vulnerability as a trans person, despite the privileges of safety afforded by my whiteness and perceived cis masculinity. I lost myself in their experiences, jumping at sights of shadows on dark streets in Montreal and feeling my throat tighten whenever I tried to write the results and analysis chapters of my Master's thesis. I became a vessel for my participants' stories, my body containing boundless rage and heartbreak for my trans siblings. I did not (and could not) prepare for the emotional challenges of gathering data that was so disturbing in its tales of violence, and my attempts at self-care were only minimally helpful. My transness linked me to my participants in a way that intimately connected their experiences of violence and transphobia to my life on the outside. Our embodied transness danced between us through the letters of our correspondence, and to this day I have not been able to completely distance the nauseating pain of the incessant, dehumanizing violence participants experienced from my life.

In spite of these emotional challenges, this research was also deeply rewarding, as many participants expressed their gratitude to have an opportunity to voice their experiences. Participants shared their appreciation that someone was in touch with them who seemed to care about their current circumstances in prison, and that someone within a university setting appreciated the gravity of their conditions. These participants expressed that our communication gave them a sense of hope that things may change in the future. I have remained in touch with one participant, Jessica, who is now involved in multiple lawsuits against her state for denying her hormones. Part of the evidence she is using to bolster her case includes her questionnaire from my research, as well as the publications and presentations I have written which discuss her contributions, which she instructed me to send to her lawyer. My point here is not that my research contributed in any way to Jessica's courage in taking on her state in a legal battle for her rights as a trans woman, but rather to illustrate how the life of this research project moved beyond myself and instead has had lasting impacts on the incarcerated trans feminine individuals involved in my work, particularly in their crafting of an affirming future for themselves while still incarcerated.

Researchers who are conducting work with trans communities must inform their methodologies by respectful and safe guidelines that are founded in a critical trans politics. As discussed earlier in this chapter, research can be considerably harmful and dangerous for trans people if it is not designed with a critical framework to understand the systems of oppression and violence in which trans people are always caught. In order to do this, researchers must exercise reflexivity on their positions of power and privilege to reflect on the forms of power and violence they may not see, and become aware of them before embarking on their research. For researchers who are not trans, this means exercising reflexivity around how cisgender privilege could inform their research design, and embarking on self-education around forms of institutional and systemic transphobia that renders the state, bureaucracies and many other institutions incredibly violent for trans people. For all researchers regardless of their gendered and sexed subjectivities, conducting research with trans people necessitates the use of a critical trans politics to inform the design and implementation of their work through reading academic and activist work by trans people who are feminine and/or femme, non-binary, racialized, Indigenous, immigrant, undocumented, disabled, incarcerated and additionally marginalized to exercise research from the bottom-up and informed by intersectional frameworks. Critical trans politics must be embedded within the research design itself; otherwise, the researcher risks harming their participants and offering conclusions that are inaccurate, biased and falsely represent the trans communities with whom they worked.

Through this research, the connections and conversations that emerged between trans people offered a powerful political intervention, particularly in the face of the overwhelming incarceration of trans feminine individuals across the US (Spade 2008–2009). When trans women of color are disproportionately criminalized and incarcerated in the US and subsequently isolated from their networks of support and denied their rights to legally or medically transition, fostering a connection with trans people on the outside challenges those forms of social death wielded upon them. Through a politics of intimacy that has emerged within my research, the connections between myself and research participants arguably cultivate a future for trans people in ways that we do not often get to experience (Rosenberg 2017). I would argue that the emotional experiences of this research taught me not the lesson of my vulnerability to emotions in conducting research, but instead the value of investing a sense of intimacy into my research. Allowing intimacy to be cultivated between myself and participants goes beyond my feelings as a researcher; it offers trans people what we are so often refused: an insistence of recognition, affirmation and life. The ethos of embedding critical trans politics in social scientific research rests in breaking isolation and rebuilding ourselves as interlinked across geopolitical, spatial and emotional boundaries. Trans people in solidarity cannot be easily broken, and that message must remain central to the research we conduct.

Notes

1 This term describes those who have been assigned male at birth and identify along a feminine gender spectrum. As a term, "trans feminine" allows for the inclusion of genderqueer and non-binary identities, as well as the identities of trans women and male-to-female transsexuals, all of whom participated in my research.

2 The majority of correctional facilities in the US place people in facilities based on their assigned sex at birth. This is often left to the discretion of individual facilities, which may place someone in a men's correctional center even if she/they have had all of her/their legal documentation changed to say female and have had forms of gender affirming surgery. Correctional facilities include state prisons, pre-transitional centers, state correctional institutions, federal correctional complexes, county jails, and US penitentiaries.

3 Those who are currently incarcerated and use the PrisCoPro's services.

4 One research participant was incarcerated in Pennsylvania, so there was possible overlap between this respondent and the Hearts on a Wire survey.

References

Browne, K., and Nash, C. J. (2010). Queer methods and methodologies: An introduction. In C. J. Nash and K. Browne, eds., *Queer methods and methodologies: Intersecting queer theories and social science research*. Farnham and Burlington: Ashgate, 1–24.

Detamore, M. (2010). Queer(y)ing the ethics of research methods: Toward a politics of intimacy in researcher/researched relations. In C. J. Nash and K. Browne, eds., *Queer methods and methodologies: Intersecting queer theories and social science research*. Farnham and Burlington: Ashgate, 167–182.

Gorman-Murray, A., Johnston, L., and Waitt, G. (2010). Queer(ing) communication in research relationships: A conversation about subjectivities, methodologies and ethics. In C. J. Nash and K. Browne, eds., *Queer methods and methodologies: Intersecting queer theories and social science research*. Farnham and Burlington: Ashgate, 97–112.

Hagner, D. (2010). Fighting for our lives: The D.C. Trans Coalition's campaign for humane treatment of transgender inmates in District of Columbia correctional facilities. *Georgetown Journal of Gender and the Law*, 11, 837–867.

Hearts on a Wire Collective. (2011). *This is a prison, glitter is not allowed: Experiences of trans and gender variant people in Pennsylvania's prison systems*. Philadelphia, PA: Emmer, Lowe, and Marshall.

Maruri, S. (2011). Hormone therapy for inmates: A metonym for transgender rights. *Cornell Journal of Law and Public Policy*, 20, 807–832.

Meth, P. (2003). Entries and omissions: Using solicited diaries in geographical research. *Area*, 35(2), 195–205.

Moran, D., Pallot, J., and Piacentini, L. (2009). Lipstick, lace, and longing: Constructions of femininity inside a Russian prison. *Environment and Planning D: Society and Space*, 27, 700–720.

Nash, C. J. (2010). Queer conversations: Old-time lesbians, transmen and the politics of queer research. In C. J. Nash and K. Browne, eds., *Queer methods and methodologies: Intersecting queer theories and social science research*. Farnham and Burlington: Ashgate, 129–142.

Peek, C. (2004). Breaking out of the prison hierarchy: Transgender prisoners, rape, and the eighth amendment. *Santa Clara Law Review*, 44(4), 1211–1248.

Probyn, E. (2011). Writing shame. In M. Gregg and G. Siegworth, eds., *The affect theory reader*. Durham & London: Duke University Press, 71–92.

Rager, K. B. (2005). Self-care and the qualitative researcher: When collecting data can break your heart. *Research News and Comment*, 23–27.

Rose, G. (1997). Situating knowledges: Positionality, reflexivities and other tactics. *Progress in Human Geography*, 12(3), 305–320.

Rosenberg, R. D. (2017). When temporal orbits collide: Embodied trans temporalities in U.S. prisons. *Somatechnics*, 7(1), 74–94.

Rosenberg, R. D., and Oswin, N. (2015). Trans embodiment in carceral space: Hypermasculinity and the U.S. Prison Industrial Complex. *Gender, Place & Culture*, 22(9), 1269–1286.

Sexton, L., Jenness, V., and Sumner, J. M. (2010). Where the margins meet: A demographic assessment of transgender inmates in men's prisons. *Justice Quarterly*, 27(6), 835–866.

Spade, D. (2008–2009). Keynote address: Trans law & politics on a neoliberal landscape. 2–27. http://zinelibrary.info/files/TransLawPolitics.pdf

Spade, D. (2011). *Normal life: Administrative violence, critical trans politics, and the limits of law*. Brooklyn, NY: South End Press.

Stanley, E. A., and Smith, N., eds. (2011). *Captive genders: Trans embodiment and the prison industrial complex*. Edinburgh, Oakland, and Baltimore: AK Press.

Sultana, F. (2007). Reflexivity, positionality and participatory ethics: Negotiating fieldwork dilemmas in international research. *ACME: An International E-Journal for Critical Geographies*, 6(3), 374–385.

Sylvia Rivera Law Project. (2007). *"It's war in here:" A report on the treatment of transgender and intersex people in New York State's men's prisons*. New York, NY: Sylvia Rivera Law Project.

Taylor, Y. (2010). The 'outness' of queer: Class and sexual intersections. In C. J. Nash and K. Browne, eds., *Queer methods and methodologies: Intersecting queer theories and social science research*. Farnham and Burlington: Ashgate, 85–96.

7 Constructing an Ethics of Depathologization

Epistemological, Methodological and Ethical Reflections in Trans and Intersex Studies

Amets Suess Schwend

Introduction

As a trans[1] and non-binary scholar, activist and artist, I made the double experience of "being researched as trans and researching trans". As an intersex[2] ally, I reviewed critically my own legitimacy of writing on intersex issues. These experiences caused me a deep feeling of ethical impasse and motivated me to rethink epistemological, methodological and ethical aspects in research practice.

When participating as an interviewee in qualitative or ethnographic research projects, more than once I made ambiguous experiences. On the one hand, I experienced a comfortable interview situation that allowed me to reflect on my trajectory and produce a narrative on my experience as a scholar, activist and artist working on human rights, depathologization and gender non-binarism. On the other hand, I felt a loss of control in the process of analysis, writing and publication, finding myself exposed to a pathologizing interpretation of my experience, misunderstandings, lack of recognition of my current gender expression and insufficient confidentiality protection (Suess Schwend 2011a, 2011b, 2016a, 2022ip).

In the role as researcher, once a trans person responded to my question if I could interview her: "I don't mind, they already have examined me so many times" (Suess Schwend 2016a: 47).[3] The identification of the interview situation with the clinical setting, and the researcher's role with the figure of medical authority, produced in me a deep ethical crisis (Suess Schwend 2011b, 2016a, 2022ip).

Frequently, in academic events I observed a split between the recognition as an academic and the identification as an activist or artist (Suess Schwend 2011b, 2016a, 2022ip). When being put in the activist or artist box, my academic trajectory was often ignored. When being identified as a scholar, I was expected not to reveal my activist or artistic experience. When visibilizing my academic-activist-artistic background, this position seemed to unsettle academic audiences.

DOI: 10.4324/9781315613703-7

Finally, as an intersex ally, I continuously asked myself about the legitimacy to write on intersex issues (Suess Schwend 2020b, 2022ip), seeking for strategies to support the human rights of intersex people without reinforcing dynamics of discourse appropriation questioned from intersex academic and activist perspectives (Bastien-Charlebois 2017; Cabral Grinspan 2009a; Cabral Grinspan and Benzur 2005; Carpenter 2021[2012]; GATE 2015a; Koyama 2002; RéFRI 2021).

> As an intersex ally, I feel a special responsibility to not speak on behalf of intersex activists, but support their demands, working for a cessation of human rights violations.
>
> (Suess Schwend 2020b: 800)

Over an extended period of time, these ethical conflicts blocked my academic production. I felt I could not write anymore within a framework that contributes to dynamics of pathologization and discursive exclusion we have questioned in depathologization activism and scholarship (Suess Schwend 2011a, 2011b, 2016a, 2022ip).

I tried to resolve this conflict by means of the following strategies: (1) reinforcing my engagement in international trans depathologization activism and support to the international intersex movement; (2) reviewing the broad field of research epistemologies, methodologies and ethics developed over the last decades in the realm of social sciences; (3) looking for metatheoretical reflections in the field of trans and intersex academic and activist knowledge production; (4) contributing to the development of principles for non-pathologizing and human rights-based research practices (Bouman et al. 2017; Suess Schwend 2011a, 2011b, 2014, 2016a, 2020b, 2022ip). In my PhD thesis and other publications, I engaged in a review of epistemological, methodological and ethical reflections in social sciences and trans and intersex studies, as a continued work in progress (Suess Schwend 2011a, 2011b, 2014, 2016a, 2020b, 2022ip). In this chapter, I will develop further these previous ideas,[4] presenting principles for an "ethics of depathologization" (Suess Schwend 2020b: 807, 2020c: 55, own translation).[5] Although this book focuses on trans studies, I included a review of reflections developed in trans and intersex studies for observing a shared interest in overcoming dynamics of pathologization and epistemic injustice in research. At the same time, I would like to highlight the differentiated experiences trans and intersex people have, and the specific priorities of trans and intersex studies and movements.

The Depathologization and Human Rights Perspective

A former rather unknown term, the concept of "depathologization", used in trans activism and scholarship, spread out over the last decade, and entered in the agenda of international and regional human rights bodies, governments, professional associations and social movements. An increasing knowledge production on trans depathologization and human rights perspectives can be observed within trans studies.[6] Over the last decade, I had the opportunity to contribute theoretical reflections on the concept (Cabral Grinspan, et al. 2016; Davy et al. 2018; Suess Schwend 2010, 2011a, 2011b, 2014, 2016a, 2017a, 2017b, 2018a, 2018b, 2020a, 2020b, 2020c, 2021, 2022ip; Suess Schwend et al. 2014, 2018).

In intersex studies and activism, the human rights framework and language can be identified as a fundamental perspective (Carpenter 2015, 2016, 2018, 2021, 2021[2012]; Conferencia Regional Latinoamericana y del Caribe de Personas Intersex 2018; First African Intersex Meeting 2017; GATE 2014, 2021; GATE et al. 2019; Ghattas 2013, 2015; Intersex Asia and Asian Intersex Forum 2018; International Intersex Forum 2011, 2012, 2013, 2017; OII Europe 2014, 2017, 2019, 2021[2018]a, 2021[2018]b; RéFRI 2021). An emerging, but less frequent use of the concepts pathologization/depathologization can be observed in intersex studies and activism. These concepts have been introduced especially in relation to the demand of depathologizing intersex-related diagnostic codes in the ICD (GATE 2021; GATE, et al. 2019), the questioning of a pathologization of intersex bodies in the clinical and social context (Cabral Grinspan 2014; Carpenter 2016), the review of "pathologising connotations of the term DSD" (Monro et al. 2021: 433) and "particular social, cultural and geographic framings that contextualise the ongoing pathologisation in this arena" (Monro et al. 2021: 435), as well as in recommendations for avoiding pathologizing language use in research (RéFRI 2021).

In this sense, trans and intersex studies and activism use both human rights language and the concepts pathologization/depathologization, with differentiated priorities and meanings.

Before defining depathologization, I would like to introduce some thoughts on the term "pathologization".

> Pathologization can be understood as the conceptualization of bodily characteristics, habits, practices, gestures, people and groups of people as mentally disordered, ill, abnormal or malformed. The demand for depathologization is a response to multiple forms of pathologization of trans and intersex people in different social fields, including social, familial, educational, academic, labor, clinical and legal contexts.[endnote]
>
> (Suess Schwend 2020b: 800)

Depathologization can be defined as a theoretical–activist perspective that conceptualize sexual, gender and bodily diversity not as a mental disorder or malformation, but as a human right (Kara 2017; Suess Schwend 2016a, 2017a, 2020b, 2022ip; Suess Schwend et al. 2014; Theilen 2014).

> In this sense, the concept of depathologization refers to the questioning, the denounce and the demand of cessation of all kind of practice based on the conceptualization of sexual, gender and bodily diversity as illness, disorder or anomaly, as well as the defense of its respect, recognition and celebration in the familial, social, educational, clinic and juridical context.
>
> (Suess Schwend 2017a: 141; own translation)[7]

The international human rights framework, as established in The Yogyakarta Principles (2007) and The Yogyakarta Principles plus 10 (2017), constitutes a relevant reference point for depathologization (Carpenter 2016; Kara 2017; Suess Schwend 2016a, 2020b, 2021, 2022ip; Theilen 2014). The Yogyakarta Principles (2007), developed in 2006 by an international expert group and presented in 2007 in the UN Human Rights Council, contributes an application of international human rights law in relation to sexual orientation and gender identity. In 2017, The Yogyakarta Principles plus 10 (2017) were published, contributing additional principles, recommendations and state obligations, as well as introducing the concept of gender expression and sex characteristics. Both documents establish principles related to the human rights of trans and intersex people. Over the last years, international and regional human rights bodies included demands of the trans and intersex movements in their strategic documents and resolutions (Council of Europe 2015, 2017; European Commission 2020; European Parliament 2011, 2015a, 2015b, 2018, 2019; Hammarberg 2009; UN 2013, 2019, 2021a; UN et al. 2016; see also OII Europe 2021[2018]a, 2021[2018]b; Suess Schwend 2016a, 2021). Due to this tight reciprocal relationship, throughout the text I will use the term "depathologization and human rights perspective".

International Trans and Intersex Activism

Over the last decade, trans and intersex movements have had strong developments on their own in local, regional and international contexts, with differentiated priorities and demands aimed at defending the human rights of trans and intersex people.

International Trans Depathologization Activism

A number of authors have described the process of emergence and internationalization of the trans depathologization movement (Bento and Pelúcio 2012; Cabral Grinspan 2010, 2011, 2014, 2017; Cabral Grinspan et al. 2016; Davy et al. 2018; Missé 2010, 2012; Missé and Coll-Planas 2010; Platero 2011; Suess Schwend 2010, 2011a, 2011b, 2014, 2016a, 2017a, 2017b, 2018a, 2018b, 2020a, 2020b, 2020c, 2022ip; Suess Schwend et al. 2014; Theilen 2014; Thomas et al. 2013a).

International trans depathologization activism emerged in the scope of the parallel review of the diagnostic manuals DSM (Diagnostic and Statistical Manual of Mental Disorders) and ICD (International Classification of Diseases and Other Health Problems), as well as the SOC (Standards of Care for Gender Identity Disorders, now Standards of Care for the Health of Transsexual, Transgender, and Gender Nonconforming People). These review processes were led by their respective editors, the American Psychiatric Association (APA 2000, 2013), the World Health Organization (WHO 2019[1990], 2018) and WPATH, World Professional Association for Transgender Health, previously

Henry Benjamin International Gender Dysphoria Association (HBIGDA 2005[2001]; WPATH 2012). The review processes included the possibility of submitting proposals from civil society, without allowing participation in decision-making processes, limiting thereby a process of "democratising diagnoses" (Davy et al. 2018: 13).

In October 2007 and 2008, local trans and LGBT activist groups organized the first parallel demonstrations for trans depathologization, taking place in various European cities on the same date (Suess Schwend 2010, 2016a). In 2009, some of these groups participated in the creation of the International Network for Trans Depathologization that launched STP, International Campaign Stop Trans Pathologization (Suess Schwend 2010, 2016a). From 2009 to 2017, STP, International Campaign Stop Trans Pathologization called each year for the International Day of Action for Trans Depathologization and listed the activities for trans depathologization taking place in different world regions (STP 2017; Suess Schwend 2018a, 2020a, 2020b). In 2017, STP, International Campaign Stop Trans Pathologization counted on the endorsement of 417 groups, networks and organizations worldwide (STP 2017; Suess Schwend 2020a, 2020b). Around 250 groups and networks from different world regions participated between October 2009 and October 2017 in 795 activities in 183 different cities of Africa, Asia, Europe, Latin America, North America and Oceania, organized on the International Day of Action for Trans Depathologization and throughout the month of October (STP 2017; Suess Schwend 2020a, 2020b). International, regional and local trans activist networks organizations and groups from different world regions published reports, declarations and press releases demanding trans depathologization, and international and regional trans activist networks developed lobbying activities for trans depathologization in international and regional human rights bodies and published shared declarations.[8]

Trans depathologization activism is based on a human rights perspective and consideration of gender binarism as socially constructed and historically situated (Suess Schwend 2016a). The most relevant demands for trans depathologization activism include the removal of the diagnostic classification of gender transition processes as a mental disorder from DSM and ICD, public coverage of trans health care and a change in the trans health care model, from a psychiatric assessment process toward an informed decision-making approach. In the legal field, trans depathologization activism claims legal gender recognition without medical requirements nor those related to civil status, age or nationality. Other relevant demands are the depathologization of gender diversity in childhood, the protection from discrimination and transphobic violence, as well as the depathologization of research practices (STP 2017; Suess Schwend 2010, 2011a, 2011b, 2014, 2016a, 2017a, 2017b, 2018a, 2018b, 2020a, 2020b, 2020c, 2021, 2022ip; Suess Schwend et al. 2014, 2018).

The demand for a removal of the diagnostic classification of gender transition as a mental disorder has received the support of regional human rights bodies, such as the Council of Europe (2011, 2015) and the European Parliament

(2011, 2015a, 2015b), as well as professional associations (SOCUMES 2011; WPATH 2010).

This discursive inclusion can be placed in the broader context of a recognition of the right to protection from discrimination and violence on grounds of sexual orientation and gender identity by UN agencies and regional human rights bodies (Suess Schwend 2016a; UN 2021a). In 2016, the mandate of an International UN Expert on Sexual Orientation and Gender Identity was established (UN 2021b).

In the ICD-11, published by the WHO 2018 online and approved by the World Health Assembly on May 25, 2019, trans-related diagnostic categories were removed from the Chapter "Mental and behavioral disorders", and the code "Gender incongruence" was included in a new Chapter "Conditions related to sexual health" (WHO 2018, 2019). Trans depathologization networks acknowledged that "trans identities are formally de-psycho-pathologized in the ICD-11", while proposing next steps for "getting rid of the remaining pathologizing language and advancing towards legal depathologization and universal health coverage" (Akahatá et al. 2019: s.p.).

Regarding trans health care, the implementation of informed decision-making approaches can be observed in some countries and regions (Davy et al. 2018; Deutsch 2012, Suess Schwend 2018a, 2020a, 2020b, 2020c). Advancements in the field of legal gender recognition and protection from discrimination can be remarked, with several countries following the example of the Argentinian Gender Identity Law that in 2012 set a reference point as the first gender recognition law without medical requirements worldwide (Davy et al. 2018; Suess Schwend 2016a, 2018a, 2020a, 2020b, 2020c). At the same time, trans people continues being criminalized, persecuted, (psycho)pathologized, excluded from access to trans health care, exposed to medical requirements in the process of legal gender recognition and/or discriminated in the labor market and social context all over the world (Amnesty International 2013, 2014; CIDH 2015; FRA 2014, 2020; Winter, Diamond, et al. 2016; Winter, Settle, et al. 2016).

International Intersex Activism

The historical development of the international intersex movement has been described by intersex authors and allies (Cabral Grinspan 2009a; Cabral Grinspan and Benzur 2005; Carpenter 2016; Chase 1998; Davis 2015; Dreger 1998, 1999[1998]; GATE 2014; Ghattas 2013, 2015; Gregori 2006; Holmes 2002, 2004, 2008; Karkazis 2008; Kessler 1998; Morland 2014; Rubin 2017). From the 1980s on, intersex activism emerged in several countries by means of local and regional associations and activist groups (Chase 1998; Davis 2015; Dreger 1998, 1999[1998]; Karkazis 2008; Kessler 1998; Rubin 2017). Over the last decade, an internationalization of the intersex movement can be observed, with the creation of international and regional networks and projects, such as Brújula Intersexual; Conferencia Regional Latinoamericana y del Caribe de Personas Intersex; IHRA, Intersex Human Rights Australia; interAct; International Intersex Forum; Intersex Asia; Intersex Day Project; Intersex People of Color for Justice; Intersex Russia; Justicia Intersex; OII, Organization International Intersex; OII Europe, as well as the International Working Group for submitting proposals in the ICD revision process coordinated by GATE. Their principal demands include the cessation of surgical interventions of genital mutilation and other non-consensual treatments in intersex newborns, children and adolescents, the end of prenatal prevention practices, infanticides and non-consensual sterilization of intersex bodies, the access to clinical information and medical records, the protection from stigmatization and discrimination, the creation of support spaces, the reparation of iatrogenic harm, as well as the depathologization of conceptualizations, terminologies and visual representations in diagnostic classifications, clinical practices and research (Australia and Aotearoa/New Zealand intersex community organisations, et al. 2017; Conferencia Regional Latinoamericana y del Caribe de Personas Intersex 2018; First African Intersex Meeting 2017; GATE 2014, 2017; GATE et al. 2019; International Intersex Forum 2013, 2017 Intersex Asia and Asian Intersex Forum 2018; Intersex People of Color for Justice 2017; OII Europe 2014, 2017, 2019, 2021[2018]b).

Over the last years, some advancements regarding human rights protection of intersex children can be observed. From 2013 on, non-consensual surgical interventions on intersex newborns, children and adolescents have been condemned by various international and regional human rights bodies (Council of Europe 2017; European Parliament 2019; UN 2013, 2019; UN et al. 2016; see also OII Europe 2021[2018]a, 2021[2018]b). Currently, they are prohibited by law in some countries and regions (Carpenter 2016; Suess Schwend 2018a, 2021; UN 2019). At the same time, intersex people continue to be exposed to early surgical interventions, pathologizing diagnostic classifications and/ or discrimination, stigmatization and interphobic violence all over the world (Amnesty International 2017; FRA 2015, 2020; UN 2019).

The Academic-Activist Gap

Over the last decade, I had the opportunity to participate in international trans depathologization activism and scholarship. When I first heard about a demonstration for trans depathologization taking place, I felt I had to get into contact with trans depathologization activism, for proposing a framework of resistance to experiences of psychiatrization, pathologization, medicalization and social normativity. Finally, I had the opportunity to participate in international networks and working groups for trans depathologization, local trans activist groups, international professional associations working for trans health, academic networks in the field of trans studies and academic knowledge production related to trans depathologization and human rights (Cabral Grinspan et al. 2016; Davy et al. 2018; Suess Schwend 2010, 2011a, 2011b, 2014, 2016a, 2017a, 2017b, 2018a, 2018b, 2020a, 2020b, 2020c, 2021, 2022ip; Suess Schwend, et al. 2014, 2018).

As an intersex ally, I have been in contact with intersex activism and scholarship, participating in international meetings of trans and intersex activists, local intersex groups and academic events related to intersex studies. More recently, I form part of INIA. Intersex – New Interdisciplinary Approaches, a European project aimed at generating knowledge that supports the well being and human rights of intersex people.[9] In several publications, I included reflections on the human rights of intersex people and the development of intersex studies (Suess Schwend 2011a, 2011b, 2014, 2018a, 2020b, 2021, 2022ip).

When I first contacted trans activists, asking them for participating in an interview, I feared to lose my status as an activist by being perceived only as an academic aimed at researching "on" trans issues. These fears were not confirmed. More than once, trans activists, trusting in my activist engagement, invited me to participate in their projects. I got involved, and finally the interview never took place. Looking behind, I think this was a way of protecting themselves from being researched (Suess Schwend 2016a, 2022ip).

As an intersex ally, I feel deeply involved with the need of dismantling and condemning human rights violations in the medical context. My activist involvement and academic activity is characterized by doubts about my legitimacy to write on intersex issues, an attitude of caution for not speaking on behalf, and the urgency for supporting the human rights of intersex people (Suess Schwend 2014, 2016a, 2018a, 2020b, 2021, 2022ip).

The participation in trans and intersex activist and academic spaces reinforced my need for bridging the academic-activist gap, developing new forms of socially engaged scholarship (Suess Schwend 2016a), in the sense of an "ethics of struggle" (Davy 2008: 95) and a "third space of critical engagement" (Routledge 1996: 411). In this process, I reflected on research epistemologies, methodologies and ethics in trans and intersex studies, and the contribution of the depathologization and human rights perspective to this discussion (Suess Schwend 2011a, 2011b, 2014, 2016a, 2020b, 2022ip).

Reviewing Epistemological, Methodological and Ethical Reflections from Trans and Intersex Perspectives

While working on depathologization and human rights in the activist field, in my academic work I continued feeling blocked by ethical doubts regarding my participation in research on trans and intersex issues from a trans, non binary and intersex ally academic-activist-artistic perspective. In order to overcome the ethical impasse, I reviewed metatheoretical reflections developed over the last decades in the broader context of social sciences, including contributions from Foucault, Bourdieu, as well as poststructural, postcolonial, feminist, queer, trans and intersex perspectives (Suess Schwend 2016a). In this process, I became aware that my ethical doubts were not new or unique, but part of a broader questioning of research practices that characterized contemporary social sciences, in the sense of a "triple crisis of representation, legitimation, and praxis" (Denzin and Lincoln 2005: 19). In various theoretical perspectives, I found broad metatheoretical reflections on epistemological, methodological and ethical aspects related to power imbalances, dynamics of social injustice and the ontological impossibility of objectivity and neutrality in the research process and academic field, as well as autoethnographic and self-reflexive narratives (Suess Schwend 2011a, 2011b, 2016a, 2020b, 2022ip). These critical reflections in social sciences helped me to place my own doubts in the context of a broader discussion (Suess Schwend 2011a, 2011b, 2016a, 2020b, 2022ip).

Within this analysis of metatheoretical discourses in social sciences, I was especially interested in epistemological, methodological and ethical reflections contributed by trans and intersex studies, with the aim of analyzing the contribution of the depathologization and human rights perspective to research epistemologies, methodologies and ethics.

I reviewed their contributions having in mind the following questions: How do trans and intersex authors and allies counter the epistemological exclusion of trans and intersex people from the academic field? Which theoretical, methodological and ethical approaches do they use in order to stop human rights violations in research? Which conceptualizations and terminologies do they propose for avoiding dynamics of pathologization? How do they deal with the academic-activist gap? What does the depathologization and human rights perspective contribute to the broader discussion on research epistemologies, methodologies and ethics in social sciences?

Over the last decades, the emergence of trans studies (Stryker 2006; Whittle 2006) and intersex studies (Bastien-Charlebois 2017; Cabral Grinspan 2009a; Carpenter 2016, 2021[2012]; Holmes 2002, 2008; Monro et al. 2021; RéFRI 2021; Thomas et al. 2013b) can be observed. These discourses can be described as "broad knowledge fields produced by trans and intersex authors and allies who share, often from a double academic-activist background, a critical view on hegemonic clinical and social paradigms regarding gender and bodily diversity" (Suess Schwend 2020b: 804).

Within trans[10] and intersex[11] studies, the questioning of the diagnostic classifications and medical models is a relevant topic, as well as the demand for depathologization in the social, health care and legal context. According to Sandy Stone (2006[1987]), "[t]he clinic is a technology of inscription" (230) and "the transsexual currently occupies a position which is nowhere, which is outside the binary opposition of gendered discourse" (230). Holmes (2002) highlights that "surgeries to assign a solid sex/gender identity are interventions in which parental anxiety is treated by proxy via the body of the intersexed child" (162). More recently, Kara (2017) states that "[t]rans people have been pathologized by psycho-medical classifications and national laws for over four decades" (4). Carpenter (2018) denounces the pathologizing character of intersex-related diagnostic codes in ICD-11.

> Over time, WHO has consistently reviewed and removed pathologizing classifications and codes associated with sexual and gender minorities from the International Classification of Diseases (ICD). However, classifications

associated with intersex variations, or differences of sex development, remain pathologized. As a result, the ICD-11 facilitates, and specifies, procedures that are regarded by UN and other institutions as violating human rights.

(Carpenter 2018: 212)

Apart from analyzing critically diagnostic classifications and clinical practices, trans and intersex authors and allies developed epistemological, methodological and ethical reflections, as well as proposals for research practices based on a depathologization and human rights perspectives.

On an epistemological level, trans and intersex authors and allies review critically the academic field, denouncing research practices "on" trans and intersex people without their participation, dynamics of discursive exclusion and reduction of trans and intersex people to a testimony role (Bastien-Charlebois 2017; Bornstein 1994; Cabral Grinspan 2009a, 2009b; Cabral Grinspan and Benzur 2005; Califia 1997; Carpenter 2021[2012]; GATE 2014, 2015; Green 2006[2000]; Hale 1997; Koyama 2002; Monro et al. 2021; Namaste 2000; Radi 2019; RéFRI 2021; Stone 2006[1987]; Stryker 2006, 2006[1994]; Vergueiro 2015; Whittle 2002).

Stryker (2006[1994]) denounces that trans voices are dismissed due to a pathologizing conceptualization of trans identities.

I live daily with the consequences of medicine's definition of my identity as an emotional disorder. Through the filter of this official pathologization, the sounds that come out of my mouth can be summarily dismissed as the confused ranting of a diseased mind.

(Stryker 2006[1994]: 249)

Califia (1997) describes dynamics of objectivization in research on trans people.

[T]o be differently-gendered is to live within a discourse where other people are always investigating you, describing you, and speaking for you; and putting as much distance as possible between the expert speaker and the deviant and therefore deficient subject.

(Califa (1997): 1–2)

Namaste (2000) identifies an "erasure" (51) of trans experiences, by means of a reduction of trans people to a "figural dimension of discourse" (52), an exclusion of trans people from institutional and academic contexts, as well as institutional practices that do not take into account the needs of trans people. GATE (2015) states the consequences of excluding trans people from knowledge production.

[M]ost of us face one of the most enduring and negative consequences of trans pathologization: we are very rarely recognized as true knowledge-makers, and given formal opportunities to be agents of those changes we all want to see in the world.

(GATE 2015: s.p.)

In relation to the role of intersex people in knowledge production, Cabral Grinspan (in Cabral Grinspan and Benzur 2005) highlights the ambivalences of the testimonial voice.

> The role of intersex people as giving testimony is without doubt fundamental, but also a double-edged sword (. . .). And this is one of the darkest aspects of intersex activism: our position as enunciators, as patients, in other words, subjects of a traditionally decreased, mutilated listening.
>
> (Cabral Grinspan in: Cabral Grinspan and Benzur 2005: 298,
> own translation)[12]

In 2014, an International Working Group coordinated by GATE developed an epistemological analysis of research on intersex topics, within a report on proposals for modifying intersex-related diagnostic codes in the ICD, submitted to WHO (GATE 2014). The Working Group indicates that "[d]espite the clear evidence of a need for changes to diagnoses and treatment protocols provided by intersex activists and experts, as well as their allies in different fields, there is still strong resistance to change" (GATE 2014: 14).

They denounce that "[c]onclusions that do not meet study expectations of successful patient outcomes serve not to prompt a fundamental review of reassessment of clinical practices, but instead serve to promote further clinical research and study[endnote]" highlighting that "[t]here is no evidence of clinician consensus on the conduct of *'normalizing'* surgeries[endnote]" (GATE 2014: 15).

Trans and intersex authors and allies reflect on forms of violence in knowledge production and epistemic injustice, referring to the concept developed by Fricker (2007) (Bastien-Charlebois 2017; Carpenter 2015; Pérez 2019; Radi 2019).

Pérez (2019) highlights that "violence is performed in relation to knowledge production, circulation and recognition: denial of epistemic agency of certain subjects, unrecognized exploitation of their epistemic resources, their objectivization, among many others" (82, own translation)[13]. Radi (2019) identifies different forms of epistemic violence in the academic field.

> [A] catalogue of practices of epistemic violence, including de-qualifying and disapproving trans★ epistemic subjectivity; objectifying; canceling epistemic authority, as well as a division of intellectual labor; instrumentalization; academic extractivism; misreadings; and colonial appropriation.
>
> (Radi 2019: 52)

With reference to Fricker (2007), Bastien-Charlebois (2017) analyzes forms of epistemic injustice intersex people are exposed to.

> To consider the effect of epistemic injustice on the process of political intersex subjectivation becomes crucial for understanding the low presence of intersex voices in the public space.
>
> (Bastien-Charlebois 2017: s.p., own translation)[14]

In relation to the terms hermeneutical marginalization and testimonial injustice discussed by Fricker (2007), the author reviews their impact on intersex people, indicating that "[t]he joint effect of hermeneutical marginalization and testimonial injustice can create a deflation of credibility" (Bastien-Charlebois 2017: s.p., own translation).[15] Bastien-Charlebois (2017) stresses that "[t]he hermeneutical marginalization is decisive for the possibility of *thinking yourself* as intersex, as a social, existential or reflexive subject" (s.p., own translation).[16] Regarding the concept of testimonial injustice, Bastien-Charlebois (2017) indicates that "[t]he deficit of credibility due to testimonial injustice is combined with the unintelligibility of the experience, perspective or analysis submitted, which strongly inhibits speaking out" (s.p., own translation).[17]

Carpenter (2015) contributes an application of the concepts of intersectionality, structural violence, epistemic injustice and social justice to the situation of intersex people, highlighting the need "for intersex to be understood as a human rights and social justice issue, and not a medical issue" (15).

Trans and intersex authors and allies also question a reduction of trans and intersex trajectories to a metaphorical symbol of gender non-binarism in affirmative approaches, such as queer theory, without taking into account the broad diversity of gender expressions, trajectories and identities that include both binary and non-binary options (Holmes 2008; Namaste 2000). In this sense, Holmes (2008) denounces that "[i]ntersex persons and states have been made to function as cultural vehicles to contain and transport anxieties about sexuality and difference" (65).

Furthermore, trans and intersex authors and allies detect ethnocentric biases in some anthropological approaches, in the sense of a romantic view on third genders/sexes in non-western cultures, without being aware of specific cultural norms or situations of socio-economic precarity that may suffer people who transit between genders and/or whose bodies do not fit in socially established sex/gender models (Holmes 2004; Towle and Morgan 2002). Towle and Morgan (2002) criticize that "[t]he transgender native is portrayed not as a normal, fallible human being living within the gender constraints of his or her own society but as an appealing, exalted, transcendent being (often a hero or healer)" (477).

They question the differentiation between "'Western' (oppressive) and 'non-Western' (potentially liberatory) gender systems" (Towle and Morgan 2002: 490).

Holmes (2004) indicates that "much of the existing work on cultural systems that incorporate a 'third sex' portray simplistic visions in which societies with more than two sex/gender categories are cast as superior to those that divide the world into just two", highlighting that "to understand whether a system is more or less oppressive than another we have to understand how it treats its various members, not only its 'thirds'" (1).

At the same time, another axis of exclusion is identified: the reduced visibility and dissemination of knowledge produced by trans and intersex authors and allies from non-English speaking countries, especially from the Global

South and East (Cabral Grinspan 2009a; Radi 2019), and colonial heritage of research practices (Vergueiro 2015).

Trans and intersex authors and allies also denounce pathologizing conceptualizations, terminologies and visual representations in clinical research related to trans and intersex issues in general (Ansara and Hegarty 2012, 2014; Bastien-Charlebois 2017; Bouman et al. 2017; Cabral Grinspan 2009a, 2009b; Cabral and Benzur 2005; Holmes 2002, 2004, 2008; Monro et al. 2021; Namaste 2000; Pyne 2014; Stone 2006[1987]; Stryker 2006[1994], 2006), and specifically to gender and bodily diversity in childhood (Ansara and Hegarty 2012; Cabral Grinspan 2009a; Cabral Grinspan and Benzur 2005; GATE 2014; Holmes 2002, 2008; Langer and Martin 2004; Pyne 2014).

Ansara and Hegarty (2012) observe the use of a "[m]isgendering language" (137) and attitudes of "cisgenderism" (141)[18] in psychological literature on gender diversity in childhood, in the sense of a pathologizing language use and lack of respect of the children's name and gender, as well as forms of "coercive queering" (149) in affirmative approaches, "lumping children with self-designated gender in the categories 'queer' or 'LGBT' without attention to whether this categorization is consensual or conceptually appropriate" (150). In another paper, Ansara and Hegarty (2014) identify additional forms of "misgendering" (261) trans people of any age are exposed to, among them dynamics of "[d]egendering" (261), by means of the use of neutral pronouns, as well as the use of "[o]bjectifying biological language", such as in "postoperative male-to-female" or "biological females" (Ansara and Hegarty 2014: 261).

Intersex authors and allies criticize a conceptualization of bodies that do not fit in the socially established standard of a female or male body as "malformation" or "abnormality", as well as the reproduction of clinical images, reinforcing a pathologizing imaginary on intersex bodies (Cabral Grinspan 2009a; Cabral Grinspan and Benzur 2005; Holmes 2002). Holmes (2002) denounces that "[t]he biomedical approach may claim to provide a neutral, merely descriptive view of intersexuality, but the classification of difference as disease is not, in fact, a neutral activity; it is an ideologically loaded choice because sex and gender norms function in the service of larger political demands" (166–167).

Cabral Grinspan (2009a) indicates that "[i]ntersexuality is usually defined as a set of syndromes that produce sexed bodies marked by 'genital ambiguity'", highlighting that "[t]his definition limits, constantly, the opportunities and experiences of celebrating" (7, own translation).[19]

On a methodological level, trans and intersex authors question dynamics of objectivization that reduce trans and intersex people to the role of "patients" or "study objects" (Cabral Grinspan 2009a; Cabral Grinspan and Benzur 2005; Fisher and Mustanski 2014; GATE 2014, 2015; Radi 2019; Stone 2006[1987]; Stryker 2006[1994]; Suess Schwend 2014, 2016a, 2020b, 2022ip; Valentine 2007). A lack of opportunities for participation in the different phases of the research process is criticized, highlighting the impact of social inequalities that may hinder horizontal collaboration structures (Cabral 2009a; Cabral and Benzur 2005; Fisher and Mustanski 2014; Namaste 2000; Valentine 2007). Furthermore, a reproduction of gender binarism in questionnaires or qualitative interview strategies is questioned (Ansara and Hegarty 2012, 2014; Bauer 2012; FRA 2014; Namaste 2000; Platero 2014; The GenIUSS Group and Herman 2014; Valentine 2007).

Trans and intersex authors and allies express reflections on research ethics, highlighting the need of avoiding structural inequalities and power dynamics (Bouman et al. 2017; GATE 2014; Reicherzer et al. 2013), as well as a reinforcement of dominant paradigms (Platero 2014).

In research related to gender diversity in childhood, Platero (2014) expresses the ethical concern that "better knowledge of this situation may be used to essentialise gender, instead of providing arguments to better understand gender constructions in different societies, along with improving families' and children's lives, which is, after all, the goal of many professionals in this field" (163–164).

Quoting the UN Special Rapporteur on torture ("structural inequalities, such as the power imbalance between doctors and patients, exacerbated by stigma and discrimination, result in individuals from certain groups being disproportionately vulnerable to having informed consent compromised", UN 2013: 7, paragraph 29, in GATE 2014: 15), the GATE Working Group observes these dynamics also "in the production of knowledge on intersex health and wellbeing" (GATE 2014: 15).

Furthermore, trans and intersex authors and allies engage in a self-reflexive review on the double academic–activist perspective, considering this role as an opportunity, a challenge and a specific ethico–political responsibility (Davy 2008; Holmes 2008; Platero 2013; Platero and Drager 2015; Stryker 2006[1994], 2006; Suess Schwend 2011a, 2011b, 2014, 2016a, 2020b, 2022ip; Towle and Morgan 2002; Whittle 2002, 2006).

> On reflection, if I was not trans, I imagine I would have been an ordinary woman (though that is too difficult to imagine), perhaps with a teaching job, cooking meals, doing the garden, and ringing up the kids. Instead, I am part of the cultural crisis of the new millennium.
>
> (Whittle 2006: xiv)

> As much as I want to make intersexuality understood from the point of view of intersexuals themselves, at the same time I am loath to pry open their/our lives, allowing yet more voyeuristic, academic curiosity to access intersexuals' already overaccessed bodies.
>
> (Holmes 2008: 64)

These epistemological, methodological and ethical reflections coincide with my own observation of discursive exclusion of trans and intersex people from knowledge production or reduction to a testimony role, pathologizing conceptualizations and language use in research practice and academic events, binary answer options in questionnaires and a frequent lack of collaborative approaches in qualitative research (Bouman et al. 2017; Suess Schwend 2011a, 2011b, 2014, 2016a, 2020b, 2022ip). Regarding ethical aspects, I felt specially identified with the reflections on the specific ethico-political responsibility of the double academic-activist role (Suess Schwend 2016a).

In my PhD thesis, due to the contradictions inherent to doing research from an academic-activist perspective and the limitations for a collaborative approach within the PhD format, I finally decided not to do interviews, but to analyze depathologization discourses by means of a literature review and autoethnographic approach (Suess Schwend 2016a). In more recent research projects, I used again qualitative research techniques, experiencing similar doubts and ethical conflicts as in the beginning. I continue perceiving the association of qualitative interviewing with other forms of examination and the social inequalities between me as researcher and the research participants, in spite of trying to adopt a collaborative approach. I continue doubting on the legitimacy and ontological possibility of interpreting the experiences of other people, and the socially constructed character of research outcomes. Furthermore, I continue feeling the academic-activist gap in research practice. Apart from these tensions, I have additional concerns, among them the question about how to avoid the production of a "waste of knowledge" within an academic system that requires a continuous research and publication practice (Suess Schwend 2016b: 24)[20], and concerns regarding the academic freedom of authors and safety of research participants in a moment of increasing trans/interphobic discourses (Suess Schwend 2022ip).

Proposals and Recommendations from Trans and Intersex Perspectives

Trans and intersex authors and allies not only question dynamics of epistemic injustice but also develop recommendations for research practices based on a depathologization and human rights perspective. Trans and intersex studies are identified as an opportunity for building up own theoretical perspectives and achieving recognition of their specific epistemological contribution (Bastien-Charlebois 2017; Cabral Grinspan 2009a; Carpenter 2021[2012]; Monro et al. 2021; Radi 2019; Stryker 2006[1994], 2006; Whittle 2006).

In the field of trans studies, Stryker (2006) highlights the production of "new epistemological frameworks" (10).

> One important task for transgender studies is to articulate and disseminate new epistemological frameworks, and new representational practices, within which variations in the sex/gender relationship can be understood as morally neutral and representationally true, and through which antitransgender violence can be linked to other systemic forms of violence such as poverty and racism.
>
> (Stryker 2006: 10)

Radi (2019) explores obstacles "trans★ epistemology" face.

[T]he fact that trans★ epistemology is not an established field; the obstacles trans★ people find for their recognition as knowledge producers; the difficulties of access and continuity in the academy; and finally, the fact that trans★ studies as a scholarly field is strongly rooted in the United States, and its production is not translated and is scarcely circulated in the rest of the world.

(Radi 2019: 44)

Carpenter (2021[2012]) identifies intersex studies as "an interdisciplinary and growing field, frequently involving collaboration with intersex-led organisations, and also involving academics with lived experience" (s.p.).

Monro, et al. (2021) highlight the "co-constitution of knowledge" (431) as a relevant characteristic of intersex studies.

The emerging field of interdisciplinary intersex studies, therefore, can be characterised by the co-constitution of knowledge with the individuals and communities it seeks to study, as intersex activists (both academics and non-academics) are important authors in the field.

(Monro et al. 2021: 431)

According to Bastien-Charlebois (2017), the inclusion of intersex perspectives in academy requires the creation of opportunities for intersex knowledge production.

Allowing a breakthrough of intersex thinking in collective and university knowledge, that constitutes a part of political subjectivation, requires the granting of space for research and exploratory reflections.

(Bastien-Charlebois 2017: s.p., own translation)[21]

From the interest of integrating the own experience without being reduced to a testimony role, the reviewed authors developed strategies for combining theoretical and autoethnographic perspectives (Bornstein 1994; Doan 2010; Cabral Grinspan 2009a; Cabral Grinspan and Benzur 2005; Califia 1997; Spade 2003, 2006; Stryker 2006[1994], 2006; Whittle 2002). The intersectional perspective is identified as a useful approach for describing the complexity of trans trajectories within their social context (Platero 2013; Valentine 2007). Some authors combine texts and visual or performative formats, describing this combination as a specific trans style (Bornstein 1994; Stryker 2006[1994]).

[B]oth my identity and fashion are based on collage. You know – a little bit from here, a little bit form there? Sort of cut-and-paste thing. And that's the style of this book. It's a transgendered style, I suppose.

(Bornstein 1994: 3)

My idea was to perform self-consciously a queer gender rather than simply talk about it, thus embodying and enacting the concept simultaneously under discussion. I wanted the formal structure of the work to express a transgender aesthetic by replicating our abrupt, often jarring transitions between genders – challenging generic classification with the forms of my words just as my transsexuality challenges the conventions of legitimate gender and my performance in the conference room challenged the boundaries of acceptable academic discourse.

(Stryker 2006[1994]: 245)

This combination of text and graphics is also present in intersex knowledge production (Cabral Grinspan 2009a).

Besides, trans and intersex authors and allies explore new forms of combining activism and academic reflection, in form of collaborative and co-research approaches (Fisher and Mustanski 2014; Namaste 2000; Singh et al. 2013), as well as archiving techniques (Crandall and Schwartz 2015; Rawson 2014).

The reviewed literature also includes suggestions for non-pathologizing conceptualizations and terminologies (Ansara and Hegarty 2012, 2014; Cabral Grinspan 2009b; Cabral Grinspan and Benzur 2005; Carpenter 2018; Holmes 2008; Monro et al. 2021; Namaste 2000; Pyne 2014; Stone 2006[1987]; Stryker 2006[1994], 2006; Tompkins 2014). The use of "they" in English (GLSEN 2017: s.p.) and "elle/-e" (Muzio 2019: s.p.; Suess Schwend 2020c: 54) or asterisk in Spanish (Cabral Grinspan 2009b; Suess Schwend 2016a) can be mentioned as examples of an inclusive, non-binary language use. The asterisk also serves for expressing the diversity of gender expressions/identities the term "trans*" includes (Tompkins 2014: 26; Radi 2019: 43; Suess Schwend 2016a: 18). Trans and intersex authors and allies also introduce non-pathologizing concepts, such as "gender independent children" (Pyne 2014: 1) or "congenital variations of sex characteristics" instead of "disorders of sex development" (Carpenter 2018: 208). Monro, et al. (2021) explain the terminology used in an editorial on intersex studies, giving preference to "intersex" (433).

Furthermore, trans and intersex authors and allies contribute recommendations for avoiding cis/endonormative biases in research practices, developing proposals for conceptualizing sexes/genders as non-binary, complex and multidimensional constructions in quantitative and qualitative research designs (Ansara and Hegarty 2012, 2014; Bauer 2012; Carpenter 2021[2012]; FRA 2014; Namaste 2000; Platero 2014; The GenIUSS Group and Herman 2014; Valentine 2007).

On an ethical level, the use of collaborative and reflexive ethical approaches is proposed, from an understanding of research ethics as a continued process that is present in all project phases and need to be aware of the own positionality and researcher role (Davy 2008; Fisher and Mustanski 2014; Reicherzer et al. 2013).

Apart from these epistemological, methodological and ethical proposals, trans and intersex authors and allies contributed ethical principles and guidelines for non-pathologizing and human rights-based research practices.

In 1997, Hale published "Suggested Rules for Non-Transsexuals Writing about Transsexuals, Transsexuality, Transsexualism or Trans", indicating that these rules are "[s]till under construction" (s.p.). The rules include, among others, recommendations related to epistemological aspects, such as recognizing that "you are not the experts about transsexuals, transsexuality, transsexualism or trans ___. Transsexuals are", interrogating the own subject position and avoiding dynamics of exotization and misrepresentation (Hale 1997: s.p.). The rules also recommend being aware of the community-based character of conversations and avoiding representations of trans people's discourses as "monolithic or univocal" (Hale 1997: s.p.). Furthermore, Hale (1997) suggests: "Focus on: What does looking at transsexuals, transsexuality, transsexualism, or transsexual ___ tell about *yourself*, *not* what does it tell you about trans" (s.p.), "[a]sk yourself if you can travel in our trans worlds" (s.p.) and take criticism as a "*gift*" (s.p.).

Namaste (2009) contributed principles for trans research, taking as a reference the perspective of "indigenous knowledge" (23) and focusing on the collective dimension of research ethics. The author refers to the principles of "relevance", in the sense "that the knowledge produced will be useful to the people and communities under investigation" (25), "equity in partnership", meaning "that people about whom one writes have an equal say and an equal voice in all aspects of empirical research" (25), as well as "ownership", including the right of the community "to keep knowledge secret" (26).

More recently, two working groups elaborated ethical principles for the periodical Conferences of WPATH, World Professional Association for Transgender Health and EPATH, European Professional Association for Transgender Health, responding thereby to proposals expressed by WPATH/EPATH members (Bouman et al. 2017).[22] The principles aims at promoting the use of respectful and non-pathologizing conceptualizations, terminologies, visual representations and clinical approaches at the Conferences (Bouman et al. 2017).

> The use of language and clinical models that are respectful, nonpathologizing, and human rights based are critical principles of ethical contemporary trans health care; and clinical models must reflect this (e.g., approaches that can be classified as "reparative therapy" are to be avoided in all cases).
>
> (Bouman et al. 2017: 1)

In the field of intersex studies, several recommendation documents can be highlighted (Carpenter 2021[2012]; Koyama 2002; RéFRI 2021).

In 2003, Emi Koyama published "Suggested Guidelines for Non-Intersex Individuals Writing about Intersexuality and Intersex People" (s.p.), indicating that they are inspired by Jacob Hale's Rules (1997). Similar to Hale (1997), Koyama (2002) recommends researchers to recognize "that you are not the experts about intersex people, intersexuality or what it means to be intersexed; intersex people are" (1). The Guidelines highlight the importance of taking into account the lives and experiences of intersex people, without using them "merely to illustrate the social construction of binary sexes" (Koyama 2002: 1). Furthermore, they recommend being aware that the writings of intersex people are part of broader, community-based conversations, not conflating intersex with LGBT, and focusing "on what looking at intersexuality or intersex people tells you about yourself and the society, rather than what it tells you about intersex people" (Koyama 2002: 2). The Guidelines recommend recognizing the diversity of intersex people and listening to the critiques of intersex people about the own work (Koyama 2002). The document concludes with a call to action for stopping non-consensual surgeries on intersex children (Koyama 2002).

RéFRI, Réseau francophone de recherche sur l'intersexuation (2021), publishes "Recommendations for a respectful research on intersexuality" (s.p., own translation),[23] indicating as a reference Koyama (2002)'s Guidelines. The RéFRI recommendations put the focus on epistemological aspects, such as respect for pronouns and gender identities, avoidance of pathologizing and stigmatizing terms, as well as knowledge of and contact with intersex associations (RéFRI 2020). Furthermore, they recommend researchers strategies for avoiding dynamics of objectivization and exotization, proposing them to take an ally role against non-consensual medical practices and for human rights (RéFRI 2020). The recommendations invite researchers to engage in reflexivity, ask themselves for the reasons they are interested in intersex issues, find out and promote the work of intersex researchers, explore lived experiences and intersectional aspects and focus the research not only on intersex people but also on doctors and psychologists (RéFRI 2020).

Morgan Carpenter (2021[2012]) developed in 2012 recommendations for "[r]esearching intersex populations" (s.p.), published on the Intersex Human Rights Australia website and updated periodically, most recently in 2021. These recommendations focus on methodological aspects in research with intersex people, especially in quantitative research designs. Carpenter (2021[2012]) recommends that researchers review research priorities, count on community input and inform themselves about intersex issues and clinical practices before engaging in intersex-related research projects. In the field of LGBTI research, the author highlights the need of disaggregating the data by population and recognizing sex characteristics as a ground distinct from sexual orientation or gender expression/identity (Carpenter 2021[2012]: s.p.). The document gives suggestions about how to take into account intersex issues in survey questions, recommending the inclusion of "non-binary options", "multiple choice answers", "[a]n open field for gender", as well as to "[a]dd a separate question on sex characteristics" (Carpenter 2021[2012]: s.p.). Finally, Carpenter (2021[2012]) provides the following recommendations: "[e]nsure the survey is relevant and meaningful", "[e]nsure the survey is accessible" and "[r]esearch – particularly research discussion of medical histories – must be trauma-informed" (s.p.).

Contribution of the Depathologization and Human Rights Perspective to Research Epistemologies, Methodologies and Ethics

The epistemological, methodological and ethical reflections contributed by trans and intersex authors and allies, written from a depathologization and human rights perspective, can be related to thoughts and doubts raised in a broader meta-theoretical discussion in contemporary social sciences, raising the question about the specific contribution of the depathologization and human rights perspective. In my PhD thesis, as well as in other publications, I reviewed this potential contribution (Suess Schwend 2011a, 2011b, 2014, 2016a, 2020b, 2022ip).

On an epistemological level, trans and intersex authors and allies question structural inequalities and power imbalances present in knowledge production processes (Bastien-Charlebois 2017; GATE 2014; Monro et al. 2021; Reicherzer et al. 2013; Stryker 2006[1994], 2006) that can be placed within broader reflections on power dynamics in social sciences contributed from different theoretical perspectives (Suess Schwend 2016a, 2020b, 2022ip). As a specific contribution to this discussion "the depathologization perspective questions pathologizing aspects of medical conceptualizations and their impact on the recognition of trans and intersex people as subjects with rights to decisional autonomy and bodily integrity, denouncing them as forms of structural violence" (Suess Schwend 2020b: 807–808).

This demand can be related to the claims of other social movements, among them the movement of (ex)users and survivors of psychiatry (Lehmann and Stastny 2007), or the movement for bodily/functional diversity (Guzman and Platero 2012).

As I described before, a trans person I wished to interview associated the interview experience with pathologizing experiences in the clinical setting (Suess Schwend 2016a, 2022ip). This questioning of the qualitative interview technique can be related to the analogy established by Foucault (1994[1984], 1999[1966], 2009[1975]) between different forms of examination in the judicial, clinical, religious and research context (Suess Schwend 2016a, 2020b, 2022ip).

Within a broader field of self-reflexive and autoethnographic perspectives (Suess Schwend 2016a, 2020b, 2022ip), the depathologization and human rights perspective proposes a self-critical review on pathologizing aspects in the own research practice, and highlights the conflict between contributing a narrative of the own trajectory and being reduced to a testimony and "patient" role (Cabral Grinspan and Benzur 2005; Davy 2008; Holmes 2008; Platero 2013; Platero and Drager 2015; Stryker 2006[1994], 2006; Towle and Morgan 2002; Whittle 2002, 2006).

As specific methodological contributions, trans and intersex studies develop recommendations regarding non-pathologizating and non-binary terminologies (Ansara and Hegarty 2012, 2014; Bouman et al. 2017; Monro et al. 2021; Namaste 2000; Valentine 2007) and answer options beyond the binary in quantitative research techniques (Bauer 2012; Carpenter 2021[2012]; FRA 2014; The GenIUSS Group and Herman 2014).

In conclusion, I highlighted that "the depathologization and human rights perspective forms part of a broader field of reflections on research epistemology, methodology and ethics developed within social sciences and sexuality studies, contributing a specific view on forms of structural and epistemic violence by means of dynamics of pathologization and psychopathologization" (Suess Schwend 2020b: 809).

Ethics of Depathologization

From this review of epistemological, methodological and ethical reflections in trans and intersex studies and my experience as trans academic, activist and artist and intersex ally working from a depathologization and human rights perspective, I propose to construct an "ethics of depathologization" (Suess Schwend 2020b: 807, 2020c: 55), as a work-in-process concept open to further developments.

Ethics of depathologization can be understood as a research practice based on the depathologization and human rights perspective, developed in the field of trans and intersex studies, but applicable to different research topics, knowledge fields and situations of structural violence and epistemic injustice. In detail, I propose the following principles[24]:

- Conceptualizating gender and bodily diversity, as well as other forms of diversity, not as an illness, disorder or malformation, but as a human right.
- Refraining from pathologizing language, using and promoting respectful, affirmative and non-pathologizing conceptualizations and terminologies.
- Respecting and using the name and pronouns each person prefers, including non-binary options, regardless of their gender expression, sex assigned at birth and bodily characteristics.
- Abstaining from a use of pathologizing images in publications or presentations.
- Respecting, recognizing and supporting knowledge production from trans and intersex perspectives, including knowledge production from non-English speaking contexts and/or from the Global South and East, questioning and deconstructing power relationships, colonialist dynamics and geopolitical inequalities.
- Promoting collaborative research methodologies, involving the participants as active partners in all stage of the research process.
- Supporting research produced by social movements, respecting collective decision-making processes and promoting collaborations between academy and activism, based on a depathologization and human rights perspective.
- Taking into account gender and bodily diversity in quantitative and qualitative research methodologies, avoiding gender binarism and cis/endocentrism in research questions, concepts and answer options.
- Using reflexive and collaborative ethics approaches, understanding research ethics not reduced to a fulfillment of formal requirements, but as an ongoing process.
- Conducting informed consent processes in research, including exhaustive information about the project, fully free consent and opportunities to review the transcript of the own interview, modify and/or delete the answers or withdraw the participation.
- Assuring the protection of the rights of all research participants and guaranteeing their safety, taking into account potential risks of exposure to forms of institutional or direct trans/interphobic violence and discrimination.

- Reflecting on the own position in the scientific field, professional practice and personal experience with gender and bodily diversity.
- Recognizing the own limits of knowledge, doubts and challenges and visibilizing this recognition in research practices and publications.
- Supporting actively the human rights protection of trans and intersex people.
- Promoting research practices based on a depathologization and human rights perspective.

Conclusions

The emergence of trans and intersex movements and studies contributed to a change of the conceptualization of gender and bodily diversity and partial advancements in the health care and legal context. At the same time, trans and intersex people continue being exposed to human rights violations and dynamics of pathologization all over the world.

From a depathologization and human rights perspective, trans and intersex authors and allies developed reflections on research epistemologies, methodologies and ethics that can be related to a broader discussion in social sciences, contributing a specific focus on dynamics of pathologization and discursive exclusion.

Based on these reflections and my own academic-activist-artistic perspective, I proposed principles for an ethics of depathologization, as a work-in-process concept open to further developments. These principles aim at promoting research practices based on a depathologization and human rights perspective and the recognition of the contribution of trans and intersex perspectives in knowledge production processes.

Notes

1 Within this text, the term "trans" refers to people "whose gender identity or gender expression does not fully correspond to the sex assigned to them at birth" (FRA 2020: 8). I use the concept as an umbrella term for a wide diversity of gender expressions, trajectories and identities, including non-binary options, being aware of the Western precedence of the term and the existence of culturally specific forms of gender transition and transgression in different world regions.
2 Within this text, the term "intersex" is used as follows: "Intersex people are born with physical sex characteristics (such as sexual anatomy, reproductive organs, hormonal pattern and/or chromosomal patterns) that do not fit typical definitions form male or female bodies. Intersex people have many different kinds of characteristics or traits. These traits may be evident prenatally or at birth, they may emerge at puberty, or become apparent later in life. Intersex people are subjected to human rights violations because of their physical characteristics" (UN 2019: 3). Throughout the text, I also use the term "bodily diversity", in reference to a broad range of sex characteristics that do not fit in the culturally established notions of a male/female sexed body.
3 Original text in Spanish: "no me importa, ya me han examinado tantas veces" (Suess Schwend 2016a: 47).

4 In two conference papers published in 2011 I first drafted reflections on epistemological, methodological and ethical perspectives in trans and intersex studies (Suess Schwend
  2011a, 2011b). In a paper published in 2014, I looked more in detail in processes of
  discursive exclusion trans and intersex people are exposed to and collective projects of
  knowledge production in trans and intersex studies (Suess Schwend 2014). In my PhD
  thesis, I developed these ideas further, reviewing the contribution of the depathologization perspective to a broader field of metatheoretical discourses in social sciences
  (Suess Schwend 2016a). In a chapter of the SAGE Handbook for Global Sexualities
  I summarized this review, introducing the concept "ethics of depathologization" (Suess
  Schwend 2020b: 807) also mentioned in another publication on trans health care ("ética
  de despatologización", Suess Schwend 2020c: 55). In this chapter, I develop further
  ideas expressed in these previous publications, contributing principles for an ethics
  of depathologization. A Spanish version of these principles are included in a chapter
  I wrote for the anthology "Restituyendo saberes y practices de investigación: etnografía
  y feminismos", in process of being published by the editorial Peter Lang (Suess Schwend
  2022ip [in publication]).
5 Original in Spanish: "ética de la despatologización" (Suess Schwend 2020c: 55).
6 See, among others Bento and Pelúcio 2012; Cabral Grinspan 2010, 2011, 2014, 2017;
  Cabral, et al. 2016; Davy 2015; Davy et al. 2018; Kara 2017; Mas Grau 2017; Missé
  2010, 2012; Missé and Coll-Planas 2010; Platero 2011; Pyne 2014; Suess Schwend
  2010, 2011a, 2011b, 2014, 2016a, 2017a, 2017b, 2018a, 2018b, 2020a, 2020b, 2020c,
  2021, 2022ip; Suess Schwend et al. 2014, 2018; Theilen 2014; Thomas 2013a; Vergueiro
  2015; Winter, et al. 2009; Winter, Diamond, et al. 2016; Winter, Ehrensaft, et al. 2016;
  Winter, Settle, et al. 2016; Winters 2008.
7 Original in Spanish: "En este sentido, el concepto de despatologización hace referencia
  al cuestionamiento, a la denuncia y a la demanda de cese de cualquier práctica basada
  en la conceptualización de la diversidad sexual, corporal y de género como enfermedad,
  trastorno o anomalía, así como a la defensa de su respeto, reconocimiento y celebración
  en el ámbito familiar, social, educativo, clínico y jurídico" (Suess Schwend 2017a: 141).
8 A review of the documents for trans depathologization published by international,
  regional and local activist networks, organizations and groups until 2015 can be found
  in Suess Schwend (2016a). For more recent documents and shared declarations for trans
  depathologization, see, among others, Akahatá et al. 2018, 2019; GATE 2015, 2021;
  Iranti 2016; STP, International Campaign Stop Trans Pathologization 2017; TGEU 2021.
9 INIA. Intersex – New Interdisciplinary Approaches (www.intersexnew.co.uk/) is a project funded by the European Union's Horizon 2020 research and innovation program
  under the Marie Skłodowska-Curie grant agreement No. 859869.
10 See, among others Ansara and Hegarty 2012, 2014; Balzer and Hutta 2012; Bento and
  Pelúcio 2012; Bouman et al. 2017; Bornstein 1994; Cabral Grinspan 2010, 2011, 2014,
  2017; Cabral Grinspan et al. 2016; Califia 1997; Crandall and Schwartz 2015; Davy
  2008, 2015; Davy et al. 2018; Doan 2010; Fisher and Mustanski 2014; GATE 2015;
  Green 2006[2000]; Guzman and Platero 2012; Hale 1997; Iranti 2016; Johnston 2019;
  Kara 2017; Mas Grau 2017; Missé 2010, 2012; Missé and Coll-Planas 2010; Namaste
  2000; Platero 2011, 2013, 2014; Platero and Drager 2015; Pyne 2011, 2014; Radi 2019;
  Rawson 2014; Reicherzer et al. 2013; Schilt and Westbrook 2009; Singh et al. 2013;
  Spade 2003, 2006; Stone 2006[1987]; Stryker 2006[1994], 2006; Suess Schwend 2010,
  2011a, 2011b, 2014, 2016a, 2017a, 2017b, 2018a, 2018b, 2020a, 2020b, 2020c, 2021,
  2022ip; Suess Schwend et al. 2014, 2018; Theilen 2014; Thomas et al. 2013a; Tompkins
  2014; Towle and Morgan 2002; Valentine 2007; Vergueiro 2015; Whittle 2002, 2006;
  Whittle, et al. 2008; Wilchins 1997; Wilson 1997; Winter, et al. 2009; Winter, Diamond,
  et al. 2016; Winter, Ehrensaft, et al. 2016; Winter, Settle, et al. 2016; Winters 2008.
11 See, among others Bastien-Charlebois 2017; Cabral Grinspan 2009a, 2009b; Cabral and
  Benzur 2005; Carpenter 2015, 2016, 2018, 2021, 2021[2012]; Chase 1998; Davis 2015;
  Dreger 1999[1998], 1999; GATE 2014, 2017; Ghattas 2013, 2015; Grabham 2007;

Gregori 2006; Holmes 2002, 2004, 2008; Iranti 2016; Jones 2018; Karkazis 2008; Kessler 1990, 1998; Koyama 2002; Monro et al. 2017, 2019, 2021; Morland 2014; RéFRI 2021; Rubin 2017; Thomas et al. 2013b.

12 Original text in Spanish: "El rol de las personas intersex como testimoniantes es sin lugar a dudas fundamental, pero también un arma de doble filo (. . .). Y ese es uno de los aspectos más sombríos del activismo intersex: nuestra posición como enunciatarios, como pacientes, es decir, sujetos de una tradicional escucha menguada, mutilada" (Cabral Grinspan in: Cabral Grinspan and Benzur 2005: 298).

13 Original text in Spanish: "la violencia es ejercida en relación con la producción, circulación y reconocimiento del conocimiento: la negación de la agencia epistémica de ciertos sujetos, la explotación no reconocida de sus recursos epistémicos, su objetificación, entre muchas otras" (Pérez 2019: 82).

14 Original text in French: "Prendre en considération le jeu de l'injustice épistémique sur le processus de subjectivation politique intersexe se révèle crucial pour comprendre la faible présence des voix intersexes dans l'espace public" (Bastien-Charlebois 2017: s.p.)

15 Original text in French: "L'effet conjugué de la marginalization herméneutique et de l'injustice testimoniale peut créer une déflation de la crédibilité." (Bastien-Charlebois 2017: s.p.)

16 Original text in French: "La marginalisation herméneutique est déterminante dans la possibilité de *se penser* intersexe, comme sujet social, existentiel ou réflexif." (Bastien-Charlebois 2017: s.p)

17 Original text in French: "Le déficit de crédibilité dû à l'injustice testimoniale se conjugue à l'inintelligibilité de l'expérience, de la perspective ou de l'analyse soumises, ce qui inhibe fortement la prise de parole." (Bastien-Charlebois 2017: s.p.).

18 Ansara and Hegarty (2012) define "cisgenderism" as "a form of 'othering' that takes people categorized as 'transgender' as 'the effect to be explained'" (141). In the recent literature, the use of the concept "cisnormativity" can be identified (Pyne 2011: 129). According to Schilt and Westbrook (2009), *"[c]is* is the Latin prefix for 'on the same side.' It compliments *trans*, the prefix for 'across' or 'over.' 'Cisgender' replaces the terms 'nontransgender' or 'bio man/bio woman' to refer to individuals who have a match between the gender they were assigned at birth, their bodies, and their personal identity" (461). The term cis or cisgender is used in trans activism and scholarship to deconstruct the conceptualization of trans people as an exception of the norm and to question the unnamed character of this culturally established norm Intersex activism and studies use endosex to refer to "a person that was born with physical sex characteristics that match what is considered usual for binary female or male bodies by the medical field" (Monro, et al. 2021: 437).

19 Original text in Spanish: "Suele definirse a la intersexualidad como un conjunto de síndromes que producen cuerpos sexuados marcados por la 'ambigüedad genital'. (. . .) Esta definición limita, incesantemente, las oportunidades y las experiencias del celebrar" (Cabral Grinspan 2009a: 7).

20 Original text in Spanish: "dispendio de conocimientos" (Suess Schwend 2016b: 24).

21 Original text in French: "Permettre une percée de la pensée intersexe dans les savoirs collectifs et universitaires, qui constitue un pan de la subjectivation politique, nécessite l'octroi d'un espace aux recherches et aux réflexions exploratoires" (Bastien-Charlebois 2017: s.p.).

22 I had the opportunity to participate in the working group that developed the Language Guide for EPATH (Bouman, et al. 2017).

23 Original text in French: "Recommandations pour une recherche respectueuse sur l'intersexuation" (RéFRI 2021: s.p.).

24 A Spanish version of these principles are included in Suess Schwend (2022ip).

References

Akahatá, APTN, Asia Pacific Transgender Network, GATE, ILGA World, RFSL, Southern African Trans Forum (SATF), STP, International Campaign Stop Trans Pathologization, & TGEU, Transgender Europe. 2018. *Joint Statement for Depathologization and TDoR 2018*. Accessed 7th December 2021. https://tgeu.org/joint-statement-for-depathologization-and-tdor-2018/

Akahatá, APTN, GATE, ILGA World, Iranti, RFSL, STP, International Campaign Stop Trans Pathologization, TGEU, Transgender Europe, & ULTRANS. 2019. *Joint Statement on ICD-11 process for trans & gender diverse people*. Accessed 7th December 2021. https://transactivists.org/icd-11-trans-process/

Amnesty International. 2013. *Because of Who I Am. Homophobia, Transphobia and Hate Crimes in Europe*. London: AI.

Amnesty International. 2014. *The State Decides Who I Am. Lack of Legal Gender Recognition for Transgender People in Europe*. London: AI.

Amnesty International. 2017. *First, Do No Harm: Ensuring the Rights of Children with Variations of Sex Characteristics in Denmark and Germany*. London: AI.

Ansara, Y.G. & P. Hegarty. 2012. "Cisgenderism in Psychology: Pathologising and Misgendering Children from 1999 to 2008." *Psychology and Sexuality* 3(2): 137–160.

Ansara, Y.G. & P. Hegarty. 2014. "Methodologies of Misgendering: Recommendations for Reducing Cisgenderism in Psychological Research." *Feminism & Psychology* 24(2): 259–270.

APA, American Psychiatric Association. 2000. *DSM-IV-TR, Diagnostic and Statistical Manual, 4th Edition, Text Revision*. Washington, DC: APA.

APA, American Psychiatric Association. 2013. *Diagnostic and Statistical Manual of Mental Disorders, Fifth Edition, DSM-5*. Arlington: American Psychiatric Publishing.

Australia and Aotearoa/New Zealand Intersex Community Organisations and Independent Advocates. 2017. *Darlington Statement*. Accessed 7th December 2021. https://intersexday.org/wp-content/uploads/2017/03/Darlington-Statement.pdf

Balzer, C. & J. Hutta. 2012. *Transrespect versus Transphobia Worldwide – A Comparative Review of the Human-rights Situation of Gender-variant/trans People*. Berlin: TGEU, Transgender Europe.

Bastien-Charlebois, J. 2017. "Les sujets intersexes peuvent-ils (se) penser? Les empiétements de l'injustice épistémique sur le processus de subjectivation politique des personnes intersex(ué)es." *Socio* 9: 143–162.

Bauer, G.R. 2012. "Making Sure Everyone Counts: Considerations for Inclusion, Identification, and Analysis of Transgender and Transsexual Participants in Health Surveys." In *The Gender, Sex and Health Research Casebook: What a Difference Sex and Gender Make*, edited by CIHR Institute of Gender and Health, 259–267. Vancouver: CIHR Institute of Gender and Health.

Bento, B. & L. Pelúcio. 2012. "Despatologização do gênero: a politização das identidades abjetas." *Revista Estudos Feministas* 20(2): 569–581.

Bornstein, K. 1994. *Gender Outlaws. On Men, Women and the Rest of Us*. Nueva York: Routledge.

Bouman, W.P., A. Suess Schwend, J. Motmans, A. Smiley, J.D. Safer, M.B. Deutsch, et al. 2017. "Language and Trans Health." *International Journal of Transgenderism* 18(1): 1–6.

Cabral Grinspan, M. 2009a. "Presentación." In *Interdicciones. Escrituras de la intersexualidad en castellano*, edited by M. Cabral Grinspan, 5–12. Córdoba, Argentina: Anarrés Editorial.

Cabral Grinspan, M. 2009b. "Asterisco." In *Interdicciones. Escrituras de la intersexualidad en castellano*, edited by M. Cabral Grinspan, 14. Córdoba, Argentina: Anarrés Editorial.

Cabral Grinspan, M. 2010. "Autodeterminación y libertad." *Página12, Suplemento Soy* 22 October 2010.

Cabral Grinspan, M. 2011. *Dilemas de la despatologización.* CLAM, Centro Latino-Americano em Sexualidade e Direitos Humanos. Accessed 7th December 2021. www.clam.org.br/destaque/conteudo.asp?infoid=7957&sid=25

Cabral Grinspan, M. 2014. "Leyendo entre líneas. Día de Acción Mundial por la Despatologización Trans." *Página12, Suplemento Soy* 17 October 2014.

Cabral Grinspan, M. 2017. "Right Answers." *Arch Sex Behav* 46(8): 2505–2506.

Cabral Grinspan, M. & G. Benzur. 2005. "Cuando digo *intersex*. Un diálogo introductorio a la intersexualidad." *Cuadernos Pagu* 24: 283–304.

Cabral Grinspan, M., A. Suess Schwend, J. Ehrt, T.J. Seehole & J. Wong. 2016. "Removal of a Gender Incongruence of Childhood Diagnostic Category: A Human Rights Perspective." *Lancet Psychiatry* 3(5): 405–406.

Califia, P. 1997. *Sex Changes. The Politics of Transgenderism.* San Francisco: Cleis Press.

Carpenter, M. 2015. *Intersex: Intersectionality, Epistemic and Structural Violence.* Accessed 7th December 2021. https://morgancarpenter.com/intersectionality-epistemic-structural/

Carpenter, M. 2016. "The Human Rights of Intersex People: Addressing Harmful Practices and Rhetoric of Change." *Reproductive Health Matter* 24(47): 74–84.

Carpenter, M. 2018. "Intersex Variations, Human Rights, and the International Classification of Diseases." *Health and Human Rights Journal* 20(2): 205–214.

Carpenter, M. 2021. "Intersex Human Rights, Sexual Orientation, Gender Identity, Sex Characteristics and the Yogyakarta Principles plus 10." *Culture, Health & Sexuality* 23(4): 516–532.

Carpenter, M. 2021[2012]. *Researching intersex populations.* Altona, Victoria, Australia: IHRA, Intersex Human Rights Australia. Accessed 7th December 2021. https://ihra.org.au/research/

Chase, C. 1998. "Hermaphrodites with Attitude: Mapping the Emergence of Intersex Political Activism." *GLQ: A Journal of Lesbian and Gay Studies* 3(2): 189–211.

CIDH, Comisión Interamericana de Derechos Humanos. 2015. *Violencia Contra Personas Lesbianas, Gays, Bisexuales, Trans e Intersex en América.* Washington, DC: CIDH.

Conferencia Regional Latinoamericana y del Caribe de Personas Intersex. 2018. *Declaración de San José de Costa Rica.* Accessed 7th December 2021. https://brujulaintersexual.org/2018/04/02/declaracion-de-san-jose-de-costa-rica/ English version: https://intersexday.org/en/san-jose-costa-rica-statement/

Council of Europe. 2011. *Discrimination on Grounds of Sexual Orientation and Gender Identity, 2nd Edition.* Strasbourg: Council of Europe Publishing.

Council of Europe. 2015. *Parliamentary Assembly. Discrimination against Transgender People in Europe, Resolution 2048(2015).*

Council of Europe. 2017. *Promoting the Human Rights of and Eliminating Discrimination against Intersex People Resolution 2191(2017).*

Crandall, M. & S.W. Schwartz. 2015. "Moving Transgender Histories. Sean Dorsey's Trans Archival Practice." *TSQ, Transgender Studies Quarterly* 2(4): 565–577.

Davis, G. 2015. *Contesting Intersex. Biopolitics: Medicine, Technoscience, and Health in the 21st Century.* New York and London: New York University Press.

Davy, Z. 2008. Transsexual recognition: Embodiment, bodily aesthetics and the medicolegal system [PhD Thesis]. Lincoln: University of Lincoln.

Davy, Z. 2015. "The DSM-5 and the Politics of Diagnosing Transpeople." *Archives of Sexual Behavior* 44(5): 1165–1176.

Davy, Z., A. Sørlie, A. & A. Suess Schwend. 2018. "Democratising Diagnoses? The Role of the Depathologization Perspective in Constructing Corporeal Trans Citizenship." *Critical Social Policy* 38(1): 5–12.

Denzin, N.K. & Y.S. Lincoln. 2005. "Introduction. The Discipline and Practice of Qualitative Research." In *The Sage Handbook of Qualitative Research. Third Edition*, edited by N.K. Denzin & Y.S. Lincoln, 1–32. Thousand Oaks, London, New Delhi: Sage Publications.

Deutsch, M.B. 2012. "Use of the Informed Consent model in the Provision of Cross-Sex Hormone Therapy: A Survey of the Practices of Selected Clinics." *International Journal of Transgenderism* 13(3): 140–146.

Doan, P. 2010. "The Tyranny of Gendered Spaces: Living beyond the Gender Dichotomy." *Gender, Place and Culture: A Journal of Feminist Geography* 17(5): 635–654.

Dreger, A.D. 1998. *Hermaphrodites and the Medical Invention of Sex*. Cambridge and London: Harvard University Press.

Dreger, A.D. 1999[1998]. "A History of intersex: From the Age of Gonads to the Age of Consent." In *Intersex in the Age of Ethics*, edited by A.D. Dreger, 5–22. Hagerstown: University Publishing Group.

European Commission. 2020. *LGBTIQ Equality Strategy 2020–2025. COM(2020) 698*.

European Parliament. 2011. *Resolución del Parlamento Europeo, de 28 de septiembre de 2011, sobre derechos humanos, orientación sexual e identidad de género en las Naciones Unidas*.

European Parliament. 2015a. *Report on the situation of fundamental rights in the European Union. A8–0230/2015, 22 July 2015*.

European Parliament. 2015b. *European Parliament resolution of 8 September 2015 on the situation of fundamental rights in the European Union (2013–214)*.

European Parliament. 2018. *Report on the situation of fundamental rights in the European Union. A8–0025/2018, 13 February 2018*.

European Parliament. 2019. *European Parliament resolution of 14 February 2019 on the rights of intersex people (2018/2878(RSP))*.

First African Intersex Meeting. 2017. *Public Statement by the African Intersex Movement*. Accessed 7th December 2021. https://intersexday.org/en/statement-african-forum-2017/

Fisher, C.B. & B. Mustanski. 2014. "Reducing Health Disparities and Enhancing the Responsible Conduct of Research Involving LGBT Youth." *Hastings Center Report* 44(s4): S28– 31.

Foucault, M. 1994[1984]. *Estética, ética y hermenéutica*. Barcelona, Buenos Aires: Paidós Ibérica.

Foucault, M. 1999[1966]. *El nacimiento de la clínica. Una arqueología de la mirada médica*. Madrid: Siglo XXI.

Foucault, M. 2009[1975]. *Vigilar y castigar*. Madrid: Siglo XXI de España Editores.

FRA, European Union Agency for Fundamental Rights. 2014. *Being Trans in the EU – Comparative Analysis of EU LGBT Survey Data*. Luxembourg: FRA.

FRA, European Union Agency for Fundamental Rights. 2015. *The Fundamental Rights Situation of Intersex People*. Luxembourg: FRA.

FRA, European Union Agency for Fundamental Rights. 2020. *A Long Way to go For LGBTI Equality*. Luxembourg: FRA.

Fricker, M. 2007. *Epistemic Injustice: Power and the Ethics of Knowing*. Oxford: Oxford University Press.

GATE. 2014. *Intersex Issues in the International Classification of Diseases: A Revision*. Accessed 7th December 2021. https://gate.ngo/intersex-issues-in-the-international-classification-of-diseases-icd/

GATE. 2015. *We Are Unstoppable! GATE Statement on the International Day of Action for Trans* Depathologization*, 24 October 2015. Accessed 7th December 2021. https://gate.ngo/we-are-unstoppable-gate-statement-on-the-international-day-of-action-for-trans-depathologization/

118 *Amets Suess Schwend*

GATE. 2017. *Submission by GATE to the World Health Organization: Intersex codes in the International Classification of Diseases (ICD) 11 Beta Draft.* Accessed 7th December 2021. https://gate.ngo/wp-content/uploads/2020/03/GATE-ICD-intersex-submission.pdf

GATE. 2021. *Depathologization.* Accessed 7th December 2021: https://gate.ngo/programs/depathologization

GATE, IHRA, Justicia Intersex, et al. 2019. *Joint Statement on the International Classification of Diseases 11.* Accessed 7th December 2021. https://ihra.org.au/35299/joint-statement-icd-11/

Ghattas, D.C. 2013. *Human Rights between the Sexes. A Preliminary Study on the Life Situations of Inter* Individuals.* Berlin: Heinrich Böll Stiftung.

Ghattas, D.C. 2015. *Standing Up for the Rights of Intersex People – How Can You Help?* Brussels: ILGA-Europe, OII Europe.

GLSEN, Gay, Lesbian, Straight Education Network. 2017. *Pronouns: A Resource.* Accessed 7th December 2021. www.glsen.org/sites/default/files/GLSEN%20Pronouns%20Resource.pdf

Green, J. 2006[2000]. "Look! No, Don't! The Visibility Dilemma for Transsexual Men." In *The Transgender Studies Reader*, edited by S. Stryker & S. Whittle, 499–508. New York: Routledge.

Gregori Flor, N. 2006. "Los cuerpos ficticios de la biomedicina. El proceso de construcción del género en los protocolos médicos de asignación de sexo en bebés intersexuales." *AIBR, Revista de Antropología Iberoamericana* 1(1): 103–124.

Guzman, P. & L. Platero. 2012. "*Passing*, enmascaramiento y estrategias identitarias: diversidades funcionales y sexualidades no-normativas." In *Intersecciones: cuerpos y sexualidades en la encrucijada*, edited by L. Platero, 125–258. Barcelona: Edicions Bellaterra.

Hale, J. 1997. *Suggested Rules for Non-Transsexuals Writing about Transsexuals, Transsexuality, Transsexualism, or trans*.* Accessed 7th December 2021. https://sandystone.com/hale.rules.html

Hammarberg, T., Council of Europe Human Rights Commissioner. 2009. *Issue Paper Human Rights and Gender Identity.* CommDH/IssuePaper(2009)2.

HBIGDA, Henry Benjamin International Gender Dysphoria Association. 2005[2001]. *Standards of Care, Compilation v1–6.*

Holmes, M. 2002. "Rethinking the Meaning and Management of Intersexuality." *Sexualities* 5(2): 159–180.

Holmes, M. 2004. "Locating Third Sexes. Transformations." *Journal of Media, Culture & Technology* 8: 1–13.

Holmes, M. 2008. *Intersex. A Perilous Difference.* Selinsgrove: Susquehanna University Press.

International Intersex Forum. 2011. *First Ever International Intersex Forum.* Accessed 7th December 2021. https://ilga-europe.org/what-we-do/our-advocacy-work/trans-and-intersex/intersex/events/first-ever-international-intersex

International Intersex Forum. 2012. *ILGA Press Release.* Accessed 7th December 2021. https://oiieurope.org/the-second-international-intersex-forum-has-just-concluded-in-stockholm-with-an-affirmation-of-seven-key-demands-and-priorities-for-intersex-people/

International Intersex Forum. 2013. *Malta Declaration.* Accessed 7th December 2021. https://oiieurope.org/malta-declaration/

International Intersex Forum. 2017. *4th International Intersex Forum – Media Statement.* Accessed 7th December 2021. https://oiieurope.org/es/4th-international-intersex-forum-media-statement/

Intersex Asia and Asian Intersex Forum. 2018. *Statement of Intersex Asia and Asian Intersex Forum.* Accessed 7th December 2021. https://intersexday.org/en/intersex-asia-2018/

Intersex People of Color for Justice. 2017. *Intersex People of Color for Justice Statement.* Accessed 7th December 2021. https://intersexday.org/en/ipoc-2017/

Iranti. 2016. *Ending Pathological Practices against Trans and Intersex bodies in Africa.* Accessed 7th December 2021. http://iranti.org.za/wp-content/uploads/2019/04/Ending-Pathologi-cal-Practices-Against-Trans-And-Intersex-Bodies-in-Africa-Toolkit-2017.pdf

Johnston, L. 2019. *Transforming Gender, Sex, Place, and Space Geographies of Gender Variance.* Abingdon and New York: Routledge.

Jones, T. 2018. "Intersex Studies: A Systematic Review of International Health Literature." *SAGE Open*, 1–22. doi: 10.1177/2158244017745577

Kara, S. 2017. *Gender Is Not an Illness: How Pathologizing Trans People Violates International Human Rights Law.* Buenos Aires and New York: GATE.

Karkazis, K. 2008. *Fixing Sex. Intersex, Medical Authority, and Lived Experience.* Durham and London: Duke University Press.

Kessler, S.J. 1990. "The Medical Construction of Gender: Case Management of Intersexed Infants." *Signs* 16(1): 3–26.

Kessler, S.J. 1998. *Lessons from the Intersexed.* New Brunswick, NJ and London: Rutgers University Press.

Koyama, E. 2002. *Suggested Guidelines for Non-Intersex Individuals Writing About Intersexuality & Intersex People.* Accessed 7th December 2021. https://isna.org/pdf/writing-guide-lines.pdf

Langer, S.J. & J.I. Martin. 2004. "How Dresses Can Make You Mentally Ill: Examining Gender Identity Disorder in Children." *Child and Adolescent Social Work Journal* 21(1): 5–23.

Lehmann, P. & P. Stastny. 2007. "Reforms or Alternatives? A Better Psychiatry or Better Alternatives?" In *Alternatives Beyond Psychiatry*, edited by P. Stastny & P. Lehmann, 402–412. Berlin: Peter Lehmann Publishing.

Mas Grau, J. 2017. "Del transexualismo a la disforia de género en el DSM. Cambios terminológicos, misma esencia patologizante." *Revista Internacional de Sociología* 75(2): 1–12.

Missé, M. 2010. "Epílogo." In *El género desordenado. Críticas en torno a la patologización de la transexualidad*, edited by M. Missé & G. Coll-Planas, 265–276 Barcelona and Madrid: Egales.

Missé, M. 2012. *Transsexualitats. Altres Mirades Possibles.* Barcelona: Col·lecció Textos del Cuerpo.

Missé, M. & G. Coll-Planas 2010. "La patologización de la transexualidad: reflexiones críticas y propuestas." *Norte de salud mental* VII(38): 44–55.

Monro, S., D. Crocetti, T. Yeadon-Lee, F. Garland & M. Travis. 2017. *Intersex, Variations of Sex Characteristics, and DSD: The Need for Change. Research Report.* Huddersfield: University of Huddersfield.

Monro, S., T. Crocetti & T. Yeadon-Lee. (2019). "Intersex/variations of sex characteristics and DSD citizenship in the UK, Italy and Switzerland." *Citizenship Studies* 23(8): 780–797.

Monro, S., M. Carpenter, D. Crocetti, G. Davis, F. Garland, D. Griffith, et al. 2021. "Intersex: Cultural and social perspectives". *Culture, Health & Sexuality* 23(4): 431–440.

Morland, I. 2014. "Intersex." *TSQ, Transgender Studies Quarterly* 1(1–2): 111–114.

Muzio, E. 2019. "En boca de todes." *Página 12, Suplemento Soy* 10 October 2019.

Namaste, V.K. 2000. *Invisible Lives. The Erasure of Transsexual and Transgendered People.* Chicago and London: University of Chicago Press.

Namaste, V.K. 2009. "Undoing Theory: The 'Transgender Question' and the Epistemic Violence." *Hypatia* 24(3): 11–32.

OII Europe. 2014. *European Intersex Meeting Riga.* Accessed 7th December 2021. https://oiieurope.org/statement-of-riga/

OII Europe. 2017. *Vienna Statement of the First European Intersex Community Event.* Accessed 7th December 2021. https://oiieurope.org/statement-1st-european-intersex-community-event-vienna-30st-31st-march-2017/

OII Europe. 2019. *Protecting Intersex People in Europe: A Toolkit for Law and Policy Makers.* Berlin: OII Europe. Accessed 7th December 2021. https://oiieurope.org/protecting-intersex-people-in-europe-a-toolkit-for-law-and-policy-makers/

OII Europe. 2021[2018]a. *List of Intersex Specific Shadow Reports to UN Committees from CoE Region and from Countries Monitoring the CoE Region.* Accessed 7th December 2021. https://oiieurope.org/list-of-intersex-specific-shadow-reports/

OII Europe. 2021[2018]b. *Intersex Resources.* Accessed 7th December 2021. https://oiieurope.org/wp-content/uploads/2018/05/International-intersex-human-rights-movement_Links-to-human-rights-documents-adressing-intersex-and-important-events_February-2021-1.pdf

Pérez, M. 2019. "Violencia epistémica: reflexiones entre lo invisible y lo ignorable." *El lugar sin límites* 1(1): 81–98.

Platero, R.L. 2011. "The Narratives of Transgender Rights Mobilization in Spain." *Sexualities* 14(5): 597–614.

Platero, R.L. 2013. *La interseccionalidad en las políticas públicas sobre la ciudadanía íntima: los discursos y la agenda política española (1995–2012)* [PhD Thesis]. Madrid: Universidad Complutense de Madrid.

Platero, R.L. 2014. "The Influence of Psychiatric and Legal Discourses on Parents of Gender-Nonconforming Children and Trans Youth in Spain." *Journal of GLBT Family Studies* 10(1–2): 145–167.

Platero, R.L. & E.H. Drager. 2015. "Two Trans★ Teachers in Madrid. Interrogating Trans★formative Pedagogies." *TSQ, Transgender Studies Quarterly.* 2(3): 447–463.

Pyne, J. 2011. "Unsuitable Bodies: Trans People and Cisnormativity in Shelter Services." *Canadian Social Work Review* 28(1): 129–137.

Pyne, J. 2014. "Gender Independent Kids: A Paradigm Shift in Approaches to Gender Nonconforming Children." *Canadian Journal of Human Sexuality* 23(1): 1–8.

Radi, B. 2019. "On Trans★ Epistemology. Critiques, Contributions, and Challenges." *TSQ, Transgender Studies Quarterly,* 6(1): 43–63.

Rawson, K.J. 2014. "Archive." *TSQ, Transgender Studies Quarterly* 1(1–2): 24–26.

RéFRI, Réseau francophone de recherche sur l'intersexuation. 2020. *Recommandations pour une recherche respectueuse sur l'intersexuation.* Accessed 7th December 2021. https://refri.hypotheses.org/recommandations-aux-chercheur·e·s

Reicherzer, S., S. Shavel & J. Patton. 2013. "Examining Research Issues of Power and Privilege within a Gender-Marginalized Community." *Journal of Social, Behavioral, and Health Sciences* 7(1): 79–97.

Routledge, P. 1996. "The Third Space as Critical Engagement." *Antipode* 28(4): 399–419.

Rubin, D.A. 2017. *Intersex Matters. Biomedical Embodiment, Gender Regulation, and Transnational Activism.* Albany and New York: State University of New York.

Schilt, K. & Westbrook, L. 2009. "Doing Gender, Doing Heteronormativity: 'Gender Normals,' Transgender People, and the Social Maintenance of Heterosexuality." *Gender & Society* 23(4): 440–464.

Singh, A.A., K. Richmond & T.R. Burnes. 2013. "Feminist Participatory Action Research with Transgender Communities: Fostering the Practice of Ethical and Empowering Research Designs." *International Journal of Transgenderism* 14(3): 93–104.

SOCUMES, Sociedad Cubana Multidisciplinaria para el Estudio de la Sexualidad. 2011. "Declaración del V Congreso de Educación, Orientación y Terapia Sexual, Cuba." *Sexología y Sociedad* 17(48): 38–39.

Spade, D. 2003."Resisting Medicine, Re/modeling Gender." *Berkeley Women's Law Journal* 18: 15–37.

Spade, D. 2006. "Mutilating Gender." In *The Transgender Studies Reader,* edited by S. Stryker & S. Whittle, 315–332. New York and London: Routledge.

Stone, S. 2006[1987]. "The Empire Strikes Back: A Posttranssexual Manifesto." In *The Transgender Studies Reader*, edited by S. Stryker & S. Whittle, 221–235. New York and London: Routledge.

STP, International Campaign Stop Trans Pathologization. 2017. *Press Release: International Day of Action for Trans Depathologization 2017*. Accessed 15th January 2021 [not accessible on 1st March 2021]. www.stp2012.info/STP_Press_Release_October2017.pdf

Stryker, S. 2006[1994]. "My Words to Victor Frankenstein above the Village of Chamounix: Performing Transgender Rage." In *The Transgender Studies Reader*, edited by S. Stryker & S. Whittle, 244–256. New York and London: Routledge.

Stryker, S. 2006. "(De)Subjugated Knowledges: An Introduction to Transgender Studies." In *The Transgender Studies Reader*, edited by S. Stryker & S. Whittle, 1–18. Whittle. New York and London: Routledge.

Suess Schwend, A. 2010. "Análisis del panorama discursivo alrededor de la despatologización trans: procesos de transformación de los marcos interpretativos en diferentes campos sociales." In *El género desordenado. Críticas en torno a la patologización de la transexualidad*, edited by M. Missé & G. Coll-Planas, 29–54. Barcelona, Madrid: Egales.

Suess Schwend, A. 2011a. "Reflexiones acerca de la despatologización". In *Actas del XII Congreso de Antropología. Lugares, tiempos, memorias. La antropología ibérica en el siglo XXI*, edited by Asociación de Antropología de Castilla y León. León: Asociación de Antropología de Castilla y León.

Suess Schwend, A. 2011b. "Reflexiones sobre la despatologización". In *Actas del X Congreso Argentino de Antropología Social. La antropología interpelada: nuevas configuraciones político-culturales en América Latina*, edited by Universidad de Buenos Aires. Buenos Aires: Universidad de Buenos Aires.

Suess Schwend, A. 2014. "Cuestionamiento de dinámicas de patologización y exclusión discursiva desde perspectivas trans e intersex." *Revista de Estudios Sociales* 49: 128–143.

Suess Schwend, A. 2016a. *'Transitar entre los géneros es un derecho'. Recorridos por la perspectiva de despatologización* [PhD Thesis]. Granada: University of Granada.

Suess Schwend, A. 2016b. "Ética de la investigación en Salud Pública: responsabilidad y utilidad social en el momento actual de crisis económica" In *Ética, salud y dispendio del conocimiento. Cuadernos de la Fundació Víctor Grífols i Lucas 38*, edited by A. Segura, 24–47. Barcelona: Fundació Víctor Grífols i Lucas.

Suess Schwend, A. 2017a. "Despatologización." In *Barbarismos queer y otras esdrújulas*, edited by R.L. Platero, M. Rosón, E. Ortega, 140–151. Barcelona: Edicions Bellaterra.

Suess Schwend, A. 2017b. "Gender Diversity in Childhood: A Human Right." *Archives of Sex Behaviour* 46: 2519–2520.

Suess Schwend, A. 2018a. "Derechos de las personas trans e intersex: Revisión del marco legislativo en el contexto español desde una perspectiva de despatologización y derechos humanos." *Revista Derecho y Salud* 28(extra): 97–115.

Suess Schwend, A. 2018b. "Diversidad de género en la infancia y adolescencia desde una perspectiva de despatologización y Derechos Humanos." In *XXII Jornadas de AndAPap 2018*, edited by AndAPap, Asociación Andaluza de Pediatría de Atención Primaria, 121–139. Sevilla: AndAPap Ediciones.

Suess Schwend, A. 2020a. "Trans Health Care from a Depathologization and Human Rights Perspective." *Public Health Reviews* 41(3): 1–17.

Suess Schwend, A. 2020b. "Questioning Pathologization in Clinical Practice and Research from Trans and Intersex Perspectives." In *The SAGE Handbook of Global Sexualities, Vol. 2*, edited by Z. Davy, A. Cristina Santos, C. Bertone, R. Thoreson & S.E. Wieringa, 798–821. London: SAGE Publications.

Suess Schwend, A. 2020c. "La perspectiva de despatologización trans: ¿una aportación para enfoques de salud pública y prácticas clínicas en salud mental? Informe SESPAS

2020." ["The Trans Depathologization Perspective: A Contribution to Public Health Approaches and Clinical Practices in Mental Health? SESPAS Report 2020."] *Gac Sanit*, 34(S1): 54–60.

Suess Schwend A. 2021. "Protegiendo el derecho a la integridad corporal y a la expresión e identidad de género en la infancia y adolescencia: el marco internacional y regional de derechos humanos." In *Infancia y juventud: retos sociales y para la democracia*: 183–208, edited by A. Rodríguez García de Cortázar and M. Venegas. Valencia: Tirant Humanidades.

Suess Schwend A. 2022ip [in publication]. "Ética de la despatologización: una perspectiva en proceso de construcción." In *Restituyendo saberes y prácticas de investigación: etnografía y feminismos*, edited by C. Gregorio Gil. New York, Bern, Berlin, Brussels, Vienna, Oxford, Warsaw: Peter Lang Group AG, Serie "Researching with GEMMA".

Suess Schwend, A., K. Espineira & P. Crego Walter. 2014. "Depathologization." *TSQ, Transgender Studies Quarterly* 1(1–2): 73–77.

Suess Schwend, A., S. Winter, Z., Chiam, A. Smiley & M. Cabral Grinspan. 2018. "Depathologising Gender Diversity in Childhood in the Process of ICD Revision and Reform." *Global Public Health* 13(11): 1585–1598.

TGEU, Transgender Europe. 2021. *Topic: Depathologisation*. Accessed 7th December 2021. https://tgeu.org/tag/depathologisation/

The GenIUSS Group & J.L. Herman 2014. *Best Practices for Asking Questions to Identify Transgender and other Gender Minority Respondents on Population-based Surveys*. Los Ángeles: The Williams Institute.

Theilen, J.T. 2014. "Depathologisation of Transgenderism and International Human Rights Law." *Human Rights Law Review* 14(2): 327–342.

The Yogyakarta Principles. 2007. *Principles on the Application of International Human Rights Law in relation to Sexual Orientation and Gender Identity*. Accessed 7th December 2021. http://yogyakartaprinciples.org/wp-content/uploads/2016/08/principles_en.pdf

The Yogyakarta Principles plus 10. 2017. *Additional Principles and State Obligations on the Application of International Human Rights Law in Relation to Sexual Orientation, Gender Identity, Gender Expression and Sex Characteristics to Complement the Yogyakarta Principles*. Accessed 7th December 2021. http://yogyakartaprinciples.org/wp-content/uploads/2017/11/A5_yogyakartaWEB-2.pdf

Thomas, M.-Y., K. Espineira & A. Alessandrin, eds. 2013a. *Transidentités: Histoire d'une dépathologisation. Cahiers de la Transidentité N° 1*. Paris: L'Harmattan.

Thomas, M.-Y., K. Espineira & A. Alessandrin, eds. 2013b. *Identités Intersexes: Identités en Débat. Cahiers de la Transidentité N° 2*. Paris: L'Harmattan.

Tompkins, A. 2014. "Asterisk." *TSQ, Transgender Studies Quarterly* 1(1–2): 26–27.

Towle, E.B. & L.M. Morgan 2002. "Romancing the Transgender Native. Rethinking the Use of the 'Third Gender' Concept." *GLQ, A Journal of Lesbian and Gay Studies* 8(4): 469–497.

UN, United Nations. 2021a. *United Nations Resolutions – Sexual Orientation and Gender Identity*. Accessed 7th December 2021. www.ohchr.org/EN/Issues/Discrimination/Pages/LGBTUNResolutions.aspx

UN, United Nations. 2021b. *Independent Expert on Sexual Orientation and Gender Identity*. Accessed 7th December 2021. www.ohchr.org/en/issues/sexualorientationgender/pages/index.aspx

UN, United Nations, Human Rights Council. 2013. *Report of the Special Rapporteur on Torture and other Cruel, Inhuman or Degrading Treatment or Punishment, Juan E. Méndez*. A/HRC/22/53. February 1, 2013.

UN, United Nations, Human Rights Office of the High Commissioner. 2019. *Background Note on Human Rights Violations against Intersex People*. Accessed 7th December 2021. www.ohchr.org/EN/Issues/Discrimination/Pages/BackgroundViolationsIntersexPeople.aspx

UN, United Nations, Human Rights Office of the High Commissioner, et al. 2016. *Intersex Awareness Day – Wednesday 26 October. End Violence and Harmful Medical Practices on Intersex Children and Adults, UN and Regional Experts Urge.* Accessed 7th December 2021. https://ohchr.org/EN/NewsEvents/Pages/DisplayNews.aspx?NewsID=20739&LangID=E

Valentine, D. 2007. *Imagining Transgender. An Ethnography of a Category.* Durham and London: Duke University Press.

Vergueiro, V. 2015. *Despatologizar é descolonizar.* Accessed 7th December 2021. https://gate.ngo/es/viviane-vergueiro-despatologizar-es-descolonizar/

Whittle, S. 2002. *Respect and Equality: Transsexual and Transgender Rights.* London: Cavendish Publishing.

Whittle, S. 2006. "Foreword." In *The Transgender Studies Reader*, edited by S. Stryker & S. Whittle, xi–xvi. New York, Routledge.

Whittle, S., L. Turner, R. Combs & S. Rhodes. 2008. *Transgender Eurostudy: Legal Survey and Focus on the Transgender Experience of Health Care.* Brussels, Berlin, ILGA Europe: TGEU, Transgender Europe.

WHO, World Health Organization. 2018. *ICD-11. Geneva: WHO.* Accessed 7th December 2021. https://icd.who.int/en

WHO, World Health Organization. 2019. *World Health Assembly Update, 25 May 2019. International Statistical Classification of Diseases and Related Health Problems (ICD-11).* Accessed 7th December 2021. www.who.int/news-room/detail/25-05-2019-world-health-assembly-update

WHO, World Health Organization. 2019[1990]. *ICD-10, International Statistical Classification of Diseases and Related Health Problems, 10th Revision.* Geneva: WHO. Accessed 7th December 2021. https://icd.who.int/browse10/2019/en

Wilchins, R. A. 1997. "Gender Identity Disorder Diagnosis Harms Transsexuals." *Transgender Tapestry* 79(31): 44–45.

Wilson, K.K. 1997. "Gender as Illness: Issues of Psychiatric Classification." In *Taking Sides – Clashing Views on Controversial Issues in Sex and Gender*, edited by E.L. Paul, 31–38. Guilford: McGraw-Hill.

Winter, S., P. Chalungsooth, Y.K. The, N. Rojanalert, K. Maneerat, Y.W. Wong, et al. 2009. "Transpeople, Transprejudice and Pathologization: A Seven-country Factor Analytic Study." *International Journal of Sexual Health* 21(2): 96–118.

Winter, S., M. Diamond, J. Green, D. Karasic, T. Reed, S. Whittle & K. Wylie. 2016. "Transgender People: Health at the Margins of Society." *The Lancet* 388(10042): 390–400.

Winter, S., D. Ehrensaft, S. Pickstone-Taylor, G. De Cuypere & D. Tando. 2016. "The Psycho-medical Case against a Gender Incongruence of Childhood Diagnosis." *Lancet Psychiatry* 3(5): 404–405.

Winter, S., E. Settle, K. Wylie, S. Reisner, M. Cabral, G. Knudson & S. Baral. 2016. "Synergies in Health and Human Rights: A Call to Action to Improve Transgender Health." *The Lancet* 388(10042): 318–321.

Winters, K. 2008. *Gender Madness in American Psychiatry. Essays from the Struggle for Dignity.* Dillon, CO: GID Reform Advocates.

WPATH, World Professional Association for Transgender Health. 2010. *WPATH De-Psychopathologisation Statement.* Accessed 7th December 2021. https://amo_hub_content.s3.amazonaws.com/Association140/files/de-psychopathologisation%205-26-10%20on%20letterhead.pdf

WPATH, World Professional Association for Transgender Health. 2012. *Standards of Care for the Health of Transsexual, Transgender, and Gender Nonconforming People, 7th version.* Accessed 7th December 2021. www.wpath.org/publications/soc

8 Transnormativities

Reterritorializing Perceptions and Practice

And Pasley, Tommy Hamilton and Jaimie Veale

This chapter is part of a broader conversation around the regulation of trans people's lives – a discussion that is particularly pertinent to the healthcare sector. While our perspectives are largely informed by Aotearoa New Zealand and Pacific contexts, we situate them among global trends in gender healthcare, impressing the need to continue these critical conversations. We have divided our contributions into three sections. To begin with, we provide key examples of how we may understand gender dynamics and the formation of transnormativities (i.e., expectations around how trans people should exist), based on Deleuze and Guattari's (1987) notions of *majoritarian* and *minoritarianism*. We argue that transnormativities tend to function as a mechanism of control, limiting the degree to which trans and gender non-conforming[1] people have agency over their lives, and typically define trans experiences relative to cisgender norms. In the second section, we concentrate on the ways that invisibility and hypervisibility are products of majoritarian categorization, which determine who gets recognized, given space, erased, ignored or a combination of these. We use examples of policy, practice and standards of care to highlight how healthcare practices currently reify transnormativities. In response to this, we provide suggestions for ways that healthcare providers can collaborate with trans people to de- and re-territorialize these contexts to better serve the needs of trans people. In the third section, we emphasize the entangled nature of majoritarian categories – in particular, gender, sexuality, race, age, class, imperialism and (dis)ability – to encourage healthcare providers to embrace the inherent complexity of engaging with people, and to facilitate a better understanding of how healthcare providers can engage with trans people in practice. In particular, we draw on Deleuze and Guattari's (1987) rhizome concept to illustrate this new way of thinking about and engaging in healthcare practices that embrace multiplicities of difference. We acknowledge that many healthcare providers are making positive strides in trans healthcare and, for many, this chapter's recommendations may seem obvious. While those individuals are part of this conversation, they are not necessarily the target audience.

Resisting the temptation for universal solutions, we hope that this discussion will prompt questions and considerations around the transnormativity of 'gender affirming' healthcare. Rather than suggesting that poor practice is

DOI: 10.4324/9781315613703-8

ubiquitous, we seek to trouble systemic transnormalizing forces, right down to the ways in which gender is conceived of, to provide the means to rethink pathways to better[2] practice.

Conceptualizing Transnormativities

Beginning this chapter presents a challenging task. On the one hand, certain assumptions need to be established; meanings made intelligible. What are transnormativities? How are they formed? What is gender? On the other hand, we seek to demonstrate how the very act of making something or someone intelligible is to render it or them comprehensible against a particular arrangement of assumptions, relationships, values, and ideas; to normalize how they are understood.

Deleuze and Guattari (1987) describe such arrangements or assemblages as majoritarianisms. These assemblages foster enclaves of power – simultaneously physical (material) and ideological (discursive) – that may be accessed by those who 'qualify'. To be (cisgender)[3] male in a patriarchal society; to be white in a white supremacist society; to be bourgeois in a capitalist system; heterosexual in a heteronormative society; to have any set of relations systemically operate in one's favor. These majoritarian categories (re)produce hierarchies of privilege and disadvantage. Their intelligibility reflects a comprehension of what affects these categories have; the meanings and materialities they produce. However, it is important to understand that they do not entirely accord with any individual or group of people. Majoritarianisms operate as a set of infrastructures that are impossible to occupy completely or perpetually; that is, normativities.

Trans majoritarianism or transnormativities represent the various ways in which trans people are expected to exist: how to appear, what sort of roles to perform, what sorts of pathways to take. Essentially, how to *be* trans. While this varies across contexts, these expectations are reliably devised to (de)limit – that is, to normalize and constrain – trans lives. Before we continue, we must acknowledge that we too engage in transnormalization throughout this text by conflating all forms of gender non-conformity under the label or prefix "trans". We acknowledge that there are countless gender terms people use to describe themselves that do not necessarily coincide with the Anglo–Western concept. In Aotearoa alone, we have several te reo Māori (Māori[4] language) terms for various genders, including takatāpui, whakawahine, tangata ira tane, hinehī, hinehua, tāhine and ira tāngata (Gender Minorities Aotearoa 2015). There are also various terms from the Pacific Island that form part of Aotearoa's gender discourses (for an extensive list, see Byrne's (2015) *Blue Print for the Provision of Comprehensive Care for Trans People and Trans Communities in Asia and the Pacific*). This conflation reflects the inadequacy of language to encapsulate the variation and evolution of gender. We consciously (albeit uncomfortably) engage in this conflation because of trans' relatively open (albeit contestable) understanding of gender in the scheme of (inevitably problematic) Western, (post)colonial discourses (Stryker et al. 2008), and because of our position as pākehā (New Zealand European (white) non-Māori) researchers not wanting to colonize another culture's term or create some recolonizing pan-gender-various term.

That withstanding, under the pretext of transnormativities, institutionally and interpersonally, access to resources, bodily autonomy, equal treatment, or basic respect often become subject to the correct performance of "transness". For example, until 2011, trans people were often required to live in their gender or have counseling for a significant period of time before being allowed access to hormone treatment (WPATH SOC v.6). There remains a requirement to live in one's gender to access lower or genital reconstructive surgeries (WPATH SOC v.7: 60). Trans people have been required to establish, through psychological assessment, a certainty around their gender (even if their physical transition does not match their gender fluidity; White Hughto et al. 2015). Trans people are still often deleted (literally or effectively) from data because their gender does not fit binary gender markers (Cruz 2014; Snelgrove et al. 2012). Moreover, the assessment of trans people's performance of these standards is sometimes subject to the purview of gatekeepers within these systems, whose measure is often a matter of whether a trans person "passes"[5] as cisgender (Budge 2015; Pitts et al. 2009). Though outright rejection may be less common in contemporary practice (Chisolm-Straker et al. 2017), it is important to consider the more subtle barriers that may coerce people toward more normative performances of gender, such as only providing binary gender options on patient information forms (Winter et al. 2016); utilizing DSM-V (2013) psychological assessment protocols, which are largely incompatible with non-binary identity (White Hughto et al. 2015); or practitioner unfamiliarity with non-binary gender discourses (Reed 2016). Furthermore, it is important to recognize that there is often a large variation in the "quality" of care provided to people, largely mediated by structural inequalities, such as ethnicity (Byrne 2015), age (Siverskog 2015) and rurality (Halberstam 2005). In Aotearoa, barriers created by wealth are somewhat removed in the earlier stages of physical transition, as a result of public funding. In contexts where this is not the case, the entanglement of wealth inequality and transnormative pressures is more evident (White Hughto et al. 2015), and we certainly see these disparities in Aotearoa when it comes to access to any gender affirming surgeries (Veale et al. 2016; Wylie et al. 2016).

While many trans people fit within what could be understood as cisnormative binary gender (and we are certainly not critical of individuals whose gender is that way), the concern lies with the coercive nature of normativities (i.e., not being afforded any other performativity without significant cost), the lack of access that trans non-conformity may result in, and the hegemonies and lateral violence this can foster among trans people. To discuss transnormativities is to extend the discussion beyond disparities between cisgender and transgender people, speaking to the fact that trans and gender non-conforming people are by no means homogenous, but the normalization of (trans)gender performativity (from within and beyond trans communities) produces coercive forces that seek to limit acceptable ways of being. This is often reflected or (re)produced in trans healthcare where treatment models denote the boundaries of intelligible transition. Systems like these disadvantage those who cannot, will not, or are unwillingly coerced into conforming to transnormativities. Furthermore, it calls into question the extent to which the possibilities of gender performativity can really meet the needs of individuals in a system that offers little relief to those outside the norm.

Such hegemonies are not limited to trans people. Glick and Fiske's (1996, 1997) Ambivalent Sexism model plots the mechanisms through which much of cisnormativity is (re)produced. Benevolent Sexism entails (positive) attitudes toward traditional gender roles, while Hostile Sexism entails (negative) attitudes toward gender non-conformity. The strong positive correlation between both subscales indicates that sexist people tend to construct a dichotomy of gender performativity: "good" women and "bad" women, "good" men and "bad" men.[6] While this binary model of sexism is insufficient to describe the variation among the gender and character of people, it reflects the polarity and essentialism of gender stereotypes. Moreover, these mechanisms show relative stability in a diverse array of contexts, suggesting that patriarchal normativities are intensely pervasive (Glick et al. 2004). In New Zealand, Sibley et al. (2007) found that individuals high in Benevolent Sexism and Right Wing Authoritarianism (a measure of general conservatism; Altemeyer 1981) in the initial survey tended to be more inclined to justify rape myths six months later (as long as the rapes happened to "bad" (i.e., gender non-conforming) women). Greater investment in gender conformity tended to serve as a justification for the punishment of deviance. Gender nonconformity results in differentiated access to systems of power, ensuring that inequality is built into material and discursive structures. This may seem too simplistic, but that is because we have artificially isolated gender normativities from other majoritarianisms. A key element of majoritarianisms is that they are inherently entangled in one another: normative gender performativity inherently assumes heteronormativity, whiteness, able-bodiedness, wealth and any other contextual elements that are systematically privileged (Meadow 2017; Snorton and Haritaworn 2013; Noble 2012; Taylor, Hines, & Casey 2010).

Power, or rather access to the enclaves of agency built into normativities, is unequally distributed throughout society. While trans and cisgender people have different relationships with power structures, the similarity between the mechanisms of control in cis- and trans-normativities speaks to their entanglement with one another. For example, a "masculine" (cisgender) woman may face discrimination for not conforming to femininities, whereas a trans man may be told that he is simply a tomboy. The former relates to the "(in)appropriate" expression of gender, while the latter reflects an ontological exclusion of the person from their gender. While both cases reflect a relationship to gender norms, they often result in a distinct set of consequences and possibilities. The "masculine" cis woman could be perceived to be a lesbian, regardless of her actual sexuality, which may arrive with homophobia (Halberstam 2005).

The trans man may face a lack of facilitation from cisgender people in his life, such as access to healthcare, as a result of disbelief but, by the same token, may also be rejected by the trans community for not adequately performing masculinity (Catalano 2015).

We are all sojourners in the flux and inflexibility of gender role expectations – some of us experience more dissonance and/or resistance than others – and this should present a common point of departure when beginning conversations with patients and peers alike, because we have all asked ourselves some variation of the question "how do I want to be in the world?" We hope this fosters empathy between healthcare providers (of all genders), their trans patients, and support networks. However, it is important to recognize that, in spite of a similar compulsion to adhere to gender norms, there are material differences in the accessibility of certain gender performativities between cis and trans people. When it comes to being trans, there are multiple layers of social expectations that do not exist for cis people: not only does one have to overcome cisnormativity (i.e., adapting the gender one was assigned at birth), but one also has to wrangle with expectations around how to do "transness", according to one's context.

Similar to Ambivalent Sexism, there emerges the notions of "good" and "bad" transness. What does it mean to be trans enough? While this is contextual, transnormative histories tend to have been entangled with cisnormative expectations, which is why transnormativities often reflect aspects of binary gender norms (Lubbers 2015). Consequently, trans people are often expected to perform gender norms that do not fit their preferred becomings, molding them into normative ways of doing gender to make them intelligible to systems, even if it comes at the expense of doing gender in the ways they might desire to. It is an extortion of people who seek to do gender differently by making their access to safety, treatment and respect contingent on their adherence to dominant ideas of how one can exist (Ansara 2012). Therefore, healthcare providers must find ways to not be complicit in the normalization of trans becomings. Ensuring more equitable access to healthcare services, including the non-normative utilization of them, is vital. We provide further ideas for how providers can do this later in this chapter.

So far, we have illustrated power as a fairly static structure, which people do or do not have access to. Instead, power should be understood as contingently defined, reflecting the contextual variation of majoritarianisms (though mechanisms, such as capitalism or ambivalent sexism, may extend the generalizability of power as norms become entangled across contexts). While there is substantial evidence that intergenerational transferral of (economic) power is best predicted by heritage (Charles and Hurst 2003) and those with power are often best equipped to consolidate it (Winters 2008), the arrangements of majoritarian systems are constantly being contested, allowing us to explain how access to power is (re)negotiated. In contrast to majoritarianisms, Deleuze and Guattari (1987) describe the way individuals operate as becoming minoritarians: heterogeneous beings with agenda that emerge from their history and context. While majoritarianisms act as striated power structures, minoritarians traverse these structures, enacting a politics in relation to these norms. Deleuze and Guattari (1987) describe this form of engagement using rhizomes as a metaphor: always in the middle of becoming something new, the rhizome is constantly adapting and, even when it is cut off, it redirects and flourishes anew. This is an apt way to conceptualize gender, illustrative of its mercurial and multiplicitous nature. Though individuals engage with majoritarianisms (i.e., norms) in various ways, to understand the formation of transnormativities, we focus on how individuals may utilize contingent power dynamics to advocate for their own (or their group's) values to be recognized and valued as part of the norm; to de-territorialize the majoritarianism, opening it up to change, then re-territorializing it in ways that are intended to offer empowerment, forming a new, more traversable terrain for that person or group.

To scaffold an understanding of the way that transnormativities form and to illustrate the entangled nature of majoritarianisms, we draw on gay and lesbian communities' re-territorialization of heteronormativities, forming homonormativities (Duggan 2002), and the subsequent fallout for queer and trans individuals who did not "fit" these new standards. Brown (2009), binaohan (2014), Santos (2013) and Puar (2013, 2007), among others, discuss various versions of the way in which otherwise dominant (e.g., predominantly white, wealthy, cisgender, able-bodied) lesbians and gay men mobilized the power they possessed utilizing respectability politics[7] with the dominant heteronormative community. Neoliberal discourses, such as "freedom of choice", "equal opportunity", and marketized relationships (i.e., seeing interactions as transactions; Brown 2012), generated an appeal to commonality. Self-determination was the price paid to divert discrimination onto those who could not or would not conform to versions of homosexuality that were more palatable to heteronormative systems. Some gay and lesbian folk operationalized the "Western dream" to show that they too wanted the house, (adopted/surrogate) kids and wedding[8]– a notable shift from the anti-patriarchal politics of previous eras (Aizura 2016; Hwahng 2016). This is a manner of sexual policing, wherein a homonormative sexual performativity – the "straight-acting" gay – is required to qualify for this re-territorialized space of privilege. Moreover, this homonormativity is always already assumedly white, able-bodied, wealthy, and otherwise occupying dominant categories. In essence, what we observed here was an operationalization of whiteness, cisnormativity, ableism and wealth, under the pretense of neoliberalism, to liberate a select few from more explicit forms of queerphobia. As long as gays and lesbians remained intelligible to heteronormativities, their access to power would be less challenged. Admittedly, there were further complexities to this history. For example, hierarchies were also observed among normative communities, wherein lesbians and bisexuals had to fight for their inclusion in majoritarian structures, but we do not have the scope to address those caveats to this example of homonormalization. We are not accusing individuals of being malicious – most of the time they are merely struggling to survive, using what privilege they have to ensure their well-being – but buying into these normativities (insofar as one privileges their status) has stratified our communities. By excluding those who rank lower in the hierarchy of a subordinated people, communities reinforce the norms that subjugated them in the first place.

The various ways that normativities coalesce in healthcare for trans and gender non-conforming people are of particular interest to us. Brown (2012) highlights the importance of recognizing that (homo)normativities function differently in different contexts, while Perez (2005) has also suggested that the globalization of healthcare means that norms are often not limited by national boundaries. Therefore, it is important to pay attention to the ways normativities operate within and between different contexts. Institutions become sites of discipline, coercing individuals into ways of being that fit into broader systems of control (Foucault 1977). Furthermore, Deleuze (1992) discusses the way narratives of choice within and beyond these institutions are often used to mask the constrained nature of those choices. Pairing power with certain performativities often compels individuals to desire to seek access to these otherwise undesirable norms, despite the hardship that seeking them will produce. As well as being a poor fit for a considerable number of trans people (Reisner et al. 2016), the ways in which access to normative performativities is rendered exclusive via economic and geographic barriers, and narrow parameters of acceptability means that, for many, these pathways are not an option in the first place. When we create archetypal processes of healthcare for archetypal notions of people, we end up caring for norms, rather than caring for people.

The valorization of normative medical routes destabilizes trans communities as the (re)privileging of this conformity is commonly used to "Other" those who cannot or will not adopt transnormativities (Lubbers 2015). The metaphor of a tightrope stretched between two high-rises can be used to illustrate this. Difficult and dangerous means of accessing other spaces are created (i.e., the tightrope), while at the same time the space that people are crossing from (i.e., non-normativity) is unsafe. Power is linked to the other side (i.e., conformity) but not everyone makes it over. Some who reach the other side also find that this normative space is not compatible with who they are either.

Like homonormativities, transnormativities have become entrenched through appeals to neoliberal systems, respectability politics and compliance with cisnormative ways of being in the world (cf. gender binaries or medicalization, Irving 2012; Kunzel 2014; Matte 2014; Sekuler 2013). Again, we acknowledge that this conformity is often a product of survival, rather than malice, and reflects the terms of acceptance presented by those who act as gatekeepers. However, investment in transnormativities further valorizes (trans) gender conformity and disenfranchises those who perform their gender otherwise. The power differential in the dynamics within which transnormativities are produced means that, even for those trans people who exist happily within the boundaries of normative binary gender, transnormative gender expectations are rigid and often do not afford individuals the possibility of fluidity or deviation from linear transition pathways, should they ever feel compelled[9].

Even if we limit our considerations to binary notions of gender identity and expression, practitioners must grasp that some trans men may (desire to) enact a feminine manhood; some trans women may (desire to) enact a masculine womanhood; others may (desire to) enact a more fluid gender under these labels; many may not want the medical treatments commonly associated with binary transitions or (desire to) enact those treatments in a manner that is incoherent to a cisnormative frame; and more yet may (desire to) enact their gender in ways that are beyond a provider's comprehension.

Many individuals who do not meet transnormative expectations do not present as they desire for fear of being denied access to the treatments they seek (Vincent, 2019). For example, a person may experience affirmation of their gender expression, such as androgyneity, until they hit a certain age (e.g., puberty) and then face familial pressure to shift to a normative gender expression that may have cultural obligations or come with the threat of shame to the whole family. On a more subtle note, it may simply be the silences that occur around alternative possibilities for transition, which mean that patients are limited in their ability to consider other ways of being (Winter et al. 2016). Moreover, trans people are often held firmly to binary gender archetypes as a burden of proof (Winters 2008). This burden of proof has and continues to translate into healthcare practices that curtail or create obstacles for how trans people can exist (WPATH 2021). The enforcement of binary gender archetypes, at the expense of actively engaging with trans people on their own terms, has huge implications for effective practice and treatment, as distrust of healthcare providers can lead to non-disclosure of issues and needing to access hormones through black market sources (Silverman 2008), which present a whole host of potential health and legal issues. In psychotherapeutic settings, little can be gained from a practitioner-client relationship that is devoid of trust (Hyde et al. 2014). Researchers who do not instill trust in their participants are likely to lose access to members of the population they seek to understand and not be able to access representative samples (Martinez-San Miguel and Tobias 2016).

These issues can be even more pronounced in the case of non-binary or agender (trans)[10] people, as they inherently trouble the limitations of cis- and trans-normativities. Comparable to the reclaimed term *queer*, which troubles heteronormativity, non-binary gender operates as a non-referential identifier that exists outside of the gender binary, and agender treats gender as inapplicable to an individual. Like those who enact binary gender identities beyond the scope of cisnormativity, non-binary and agender people are often afraid to honestly engage healthcare institutions because they do not fit binary transition models (Bilodeau 2005; Joynt and Bryson 2012; Lykens 2016). Often, they feel the need to perform binary gender to attain the services or treatment they desire. The consequences of not doing so – of occupying undefined, unintelligible space – can be serious: increased experiences of harassment in school (Clark et al. 2014), elevated risk of minority stress (Meyer 2015) and postponing health care due to "fear of bias" (Lykens 2016). Understandably, many who have access to the vestige of safety through conformity do so, even if this reinforces the systems that keep them bound to transnormativities and others subject to greater inequality. However, as the term non-binary becomes more widespread in popular culture (i.e., intelligible to normative discourse via assimilation; Dowling (2017); Richard et al. (2016); Jones et al. (2016)), and therein a more possible performativity, we need to be aware of the ways in which privileges (e.g., whiteness) allow certain people to perform their non-binary gender relatively unafflicted, while others are met with resistance (Noble 2012; Snorton and Haritaworn 2013). There are objections within the trans community to the typification of non-binary folk as white, skinny, androgynous, eccentric and assigned female at birth (Finch 2015). It is imperative that healthcare providers be conscious of these implicit and explicit (White Hughto et al. 2015) biases, so that they do not reproduce them in their practice.

Visibility and Disclosure

To say that trans healthcare is transnormative is to speak to the constrained pathways that the standards of care generate within the medical industrial complex, limiting and complicating access to various medical procedures and mental health support needs. For example, psychological assessments are designed to ensure patients understand the possible ramifications of their treatment, including their certitude around their gender (Winter et al. 2016). For gender fluid or non-binary people, while physical transition may be required to embody their genders, these protocols do not speak to their experiences of gender (Byrne 2015). Also, for these same people seeking surgical intervention, what does spending a significant time performing non-binary gender entail? How does that requirement serve their decision process? Furthermore, there is little cultural sensitivity built into models (Byrne 2015; Winter et al. 2016). For example, individuals from Pacific backgrounds who fall outside of traditional Western binaries are often interpreted through binary transition models (Roen 2006).

There is a risk that people whose genders differ from the trajectory of traditionally binary Western transition may feel pressured to conform to these Western medical pathways or believe that Western transitional pathways are the only ones available (Ashleigh 2014). We acknowledge that there is limited research on non-binary or Pasifika experiences of healthcare systems, but we urge healthcare providers to consider that this may not simply be because there is no demand for gender affirming healthcare from those populations (Roen, 2006). The histories (and, in many cases, present states) of gender affirming healthcare provision are steeped in various forms of gatekeeping, such as "all or nothing" requirements, traditional gender role performativities, and pathologization (Winter et al. 2016). For those who engage with medical systems, the construction of treatment pathways becomes a matter of whether individuals are intelligible to providers, which impacts on whether patients can access the treatments they desire. For those who desire treatment outside the limited options current standards can facilitate, strategic disclosure is required to attain the best possible treatment regime to fit their desires. Because standards of care have traditionally been structured around binary gender norms and have been prohibitive of non-normative transition, this is often not possible or results in a poor fit of treatment (Reisner et al. 2016). As standards of care begin to expand their approaches to include non-binary experiences (WPATH 2021), it is important that healthcare providers incorporate these understandings into their practices, if they have not already. Furthermore, practitioners should be mindful that they do not fall into the trap of reducing non-binary people to stereotypes, as there is no discrete way to be non-binary.

However, there are numerous standards and systems that healthcare professionals may work with, around or against, which influence their capacity to facilitate different needs. Within the international human rights sector, guidance is offered to the executive and management health systems via recommendations from the World Health Organisation and the United Nations. Both these organizations promote the Yogakarta Principles, which address equitable access to healthcare. For example, Principle 17(e) states that healthcare providers are required to:

> Ensure that all people are informed and empowered to make their own decisions regarding medical treatment and care, on the basis of genuinely informed consent, without discrimination on the basis of sexual orientation or gender identity.
> (Yogyakarta Principles Plus 10, 2017 and Yogakarta Principles 2006)

Similarly, the World Professional Association of Transgender Health, Standards of Care 7 (WPATH SOC7 2011) also draws on the concepts of person-centered care, informed consent and collaboration. Numerous inclusive and affirming healthcare providers reference the WPATH SOC7 to ensure "good practice" when assisting the needs of trans people. Examples from 2016 include the New York clinic, Callen Lorde Community Health Centre, *Protocols for the Provision of Hormone Therapy* and the Center of Excellence for Transgender Care in San Fransisco *Guidelines for the Primary and Gender Affirming Care of Transgender and Gender Non Binary People* (Deutsch 2016). In 2017, the Equinox Gender Centre, based in Melbourne, Australia, run by the Victorian Aids Council, published *Protocols for the Initiation of Hormone Therapy for Trans and Gender Diverse Patients* (Cundill and Wiggins 2017). In 2018, Oliphant et al. produced the *Guidelines for Gender Affirming Healthcare for Gender Diverse and Transgender Children, Young People and Adults in Aotearoa New Zealand*. While we recognize that these provide good examples of trans-led healthcare models, we caution against the assumption that their presence is automatically met with (immediate) uptake by the medical community, that they are being practiced by a majority of medical professionals or that their practice is not colored by predispositions towards traditional Western models of gender confirmation.

Despite this, the recognition and engagement with trans health professionals, allies, community representatives and families, as well as the access to and use of these protocols and standards to inform "good practice", has offered some shifts in the trans healthcare field's recognition of trans people's needs. Examples observed informally via dialogue on social media from groups of community and health carers include the development of collective models of affirmed care or in health sector language, multi-disciplinary teams; the recognition of "no care equals harm[11]"; the introduction of informed consent forms; focus on suicide prevention; recognition of trans people's health over a life span; aged care awareness; and improved data collection, to inform future "good practice". However, while this may have reduced barriers, without critical inquiry into the ways in which discussion may be limited to issues that are already intelligible to healthcare providers, this may have reinforced transnormativities, rather than challenged limited access.

One way these "good practices" become murky can be found in the way we politely negotiate with expert health professionals. For example, as a person discloses and acknowledges the access to health care, there can be limited space to make a complaint about a health practitioner's "good practice" not meeting the person's expectations of care. Furthermore, rather than simply replacing prior foci with the issues of the day, we need to consider whether and how new foci may become the new orthodoxy. For example, while the "no care is harm" discourse is useful to provide access to those who previously struggled to get treatment, is it possible that, as the new dominant narrative, it could result in pressures to seek treatment beyond the scope of an individual's desires? While it is positive that consultation is occurring, we must accept that the opinions registered will often be those that are privileged among the previously unregistered. We risk facilitating a re-territorialization of healthcare practices that incorporates the ideas of those who are reachable, but undermine those who continue to not have their interests registered; a new norm of who is included and who is excluded. This becomes increasingly problematic for people who sit at the intersection of multiple minority statuses, generating what Purdie-Vaughns and Eibach (2008) call *intersectional invisibility* because of an individual's unidentifiability against stereotypes, which underscores the need for healthcare strategies that do not rely on models that are based on reductive categorizations of people. What strategies can ensure that healthcare engagements and understandings are facilitating the needs of all trans people?

The development of the American Psychological Association (APA) "Guidelines for Psychological Practice with Transgender and Gender Nonconforming People" (2015) highlights a cultural shift in the affirmative stance taken in supporting trans people. These guidelines explicitly state the distinction between guidelines and standards, pointing out: "Standards are mandates to which all psychologists must adhere (e.g., Ethical Principles of Psychologists and Code of Conduct; APA 2010), whereas guidelines are aspirational" (APA 2015: 833). The term "aspirational" offers an opportunity to lessen the static or fixed positioning, allows for some flexibility in the relationship between the trans person and the health worker, and affords the possibility of practices or guidelines that can assist to de-center a health professional's power.

While standards of care and guidelines may now advocate a more client-centered, flexible, and informed consent-focused approach, these developments have occurred within a context that privileges a Western perspective (Roen, 2006). This privilege is evident in the persistent lack of reference to culture in the WPATH, AusPATH, PATHA, and other documents, standards and pathways. White/colonial histories are ignored and rich indigenous gender diverse cultures are erased and overwritten by European/Western perspectives on gender diversity that only emerged in the early 1900s (White Hughto et al. 2015). Roen (2006) counters the dominant white, Western understanding of gender and sexuality, introducing the liminal space between gender and sexuality. They reminds us that, for indigenous "gender liminal" people, their identities have been made invisible by colonization's attempts to assimilate and annihilate. Trans people from non-Western cultures may engage with gender in a different language (literally and figuratively). Unfortunately, disclosure is often met with practitioners' attempts to fit non-Western cultural ideas into Western frameworks (Roen 2006). There is a need to reterritorialize these positions and identities through the reclamation of their right to identify through their cultures, if they so choose (Ashleigh 2014). As it stands, standards of care continue to embody the naturalization of Western values through medicine. In this way, transnormalization through these standards is tantamount to the recolonization of gender through medical models. This highlights the need for healthcare providers to be able to converse with patients on and in the patients' terms, and these expectations need to be written into the standards of care so that trans people from diverse cultural contexts can access the treatments they desire in the way they desire.

As we have recognized, every one of us – cis and trans alike – engage with questions of how we want to be in the world. For some trans people, their desired mode of being requires medical treatment, whether that involves hormones, cosmetics and/or surgery. Given our histories of disenfranchisement, pathologization and Othering, it is perhaps not surprising that the eventual development of medical treatment (by cisnormative institutions) has almost exclusively been built on cisnormativities. The medicalized nature of engagement has leant toward the pathologization of trans people, wherein they needed to be "fixed" (Clark et al. 2012; Snelgrove et al. 2012). In some domains, these attitudes remain (White Hughto et al. 2015). As trans people became intelligible through these and other institutions, they became part of the social order. However, operating within these systems of meaning (i.e., conforming to the trans norms that emerged) have generated hegemonies among trans people, based on the extent to which they have been able to meet these standards. It is important to acknowledge that the accessibility of these standards is inextricably tied to other forms of inequality, including economics, culture and ableism. Majoritarianisms cannot be disentangled; however, we focus on transnormative pressures because we can address aspects of these issues through improving healthcare relationships (for which there are not such simple solutions in the case of classism and other inequalities).

Notably, the pressures to conform are felt on both sides of the relationship – the carer and the cared for – as they wrestle with ideas around "good" outcomes, but it is the practices that serve to reinforce normativities, which reveal the power differential in these circumstances. Often, trans people will have an interpretation or perspective of medical protocols as barriers and the health professional as the gatekeeper. Two aspects of the health professional and patient relationship inform this position: first, that trans people "must engage with the medical system in order to modify their body", and second, that the health professional must seek a diagnosis, which leads to the pathologization of trans people (Cruz, 2014). One concern is that there is a lack of knowledge or conceptual understanding of gender. Another concern is that the healthcare professional is positioned to "fix" the pathology and takes responsibility for the trans person to not regret their choice to transition (Snelgrove et al. 2012). This is reinforced by deficit model research that centers on trans people's personal and interpersonal suffering (Reisner et al. 2016; Wylie et al. 2016), which feeds back into transnormalizing discourses (both in terms of how trans people understand their prospects, and in the biases and assumptions practitioners weave into their practice). Decentering the medical professional and allowing the transgender person to define the point of departure in their healthcare may provide a space to gain common understanding.

Rather than taking the position that these models fit most patients (a common but somewhat unsubstantiated claim; Pega and Veale 2015; Winter et al. 2016; Wylie et al. 2016), we seek to highlight the consequences of not fitting medical frameworks. Non-conformity can lead to failed attempts at access to care or a lack of sustainability in the care offered. For example, the International Classification of Diseases (ICD-11; WHO, 2018) diagnoses trans children with "gender incongruence of childhood". This diagnosis risks pathologizing children before they even engage with hormone treatments, surgeries, and cosmetic changes, if they even seek them out. This raises questions around the right to name your gender and for that self-identification to be open to change (Drescher et al., 2012; Siverskog 2015; Winter et al. 2016). We agree with the Global Action for Trans Equality (GATE) statement that "research has repeatedly affirmed that there is no way of reliably forecasting gender identity and/or gender expression in adolescence and adulthood based on gender variance in childhood" (GATE 2013). The tendency toward normalization opens practices up to the risk of health professionals "gatekeeping", even if it is unintentional, by simply adhering to new models that are uncritical of their transnormalizing nature.

As things stand, trans people's relationships with healthcare systems are not the type of relationships one can choose to (dis)engage with easily (Hwahng 2016). One's gender will always be a complex deliberation and medical treatment a serious consideration, but the authority over who determines legitimate genders and how one should perform them is a personal question, enmeshed in sociocultural contexts, rather than a medical question. This chapter confronts the ways in which the establishment of transnormativities has produced hegemonies based on degrees of conformity. Access to treatment may depend on the conformity to standards set by the medical professional, rather than addressing individuals' needs. For those who cannot or will not conform, this may result in a lack of access to treatment or being pressured into unwanted treatments, despite their desires or needs (Roberts and Fantz 2014; Vance et al. 2015).

We acknowledge that the spaces between various understandings of gender roles, expressions and assumptions can lead to the most experienced expert health professional fumbling across an unknown landscape. Practitioners may find themselves having to navigate alongside trans people whose gender is unintelligible to any model they know of. Rather than coercing patients into categories, stereotypes, and development models, we hope to broaden the thinking around the processes they employ in the pursuit of satisfactory healthcare. While we cannot address all the ways in which trans and gender non-conforming people are pathologized and normalized, we recommend that healthcare providers involved in discussions around the mitigation of these issues might like to begin by considering questions around key debates in the field.

For example, as there is a general consensus that the experiences of being trans are not a pathology (Drescher et al. 2012; Wylie et al. 2016), how can we work toward recognizing pathologizing practices and work toward depathologized forms of trans health care? In contexts where healthcare is entangled with insurance, what initial steps need to be taken to depathologize and dismantle insurance systems that rely on pathologization? In the case of the ICD-11, the introduction of z codes and the deletion of the diagnosis Gender Incongruence in Children in the ICD 11 (Drescher et al. 2012) could be one way to depathologize insurance-dependent healthcare contexts; however we should remain mindful of the ways in which positioning gender within sexual health contexts may continue to pathologize gender non-conformity.

Efforts to have trans needs reflected in the infrastructure of society have increased as trans people have become more intelligible. Setting aside the problematic nature of population surveillance, in the context of public policy that is somewhat dependent on data, there is a need to count trans people through the protocols of (state) population data collection. In 2018, Statistics New Zealand continued to restrict population to binary genders through the census. These approaches perpetuate cisnormative binaries and biological essentialist discourses that coerce trans people to be identified by their assigned gender at birth. In defense of this, Statistics New Zealand claimed that incorporating gender identity would render data too complex (as if gender and sex were simple categories to begin with; Butler 1990), thereby reifying and institutionalizing gender norms (Statistics New Zealand 2018). However, in 2023, Statistics New Zealand have committed to incorporating an 'another gender' option on the New Zealand Census (Statistics New Zealand 2021), which indicates a step in the direction of recognizing trans and non-binary people in Aotearoa New Zealand. While this provides evidence of forthcoming change, we caution the assumption that these issues have been resolved as the ways in which gender and sex are categorized remain contested.

Practitioners need to question how they frame the relationship between disclosure and consent. Gender confirmation processes that limit autonomy may jeopardize patients' engagement with these processes or cause them to disengage with care, seek non-standard treatments outside of the medical field or suffer mental health distress and other stigma-related problems, both economically and socially (Cruz 2014; Hughto et al. 2015; Poteat et al. 2013; Roberts and Fantz 2014; Winters et al. 2016; Veale et al. 2019). Discrepancies between what practitioners and clients understand as informed consent tend to reflect dynamics where power is unequally distributed. We believe it is imperative that practitioners empower their patients, ensuring they have mutual understandings. This necessarily involves critically adapting practices to promote common understandings of gender affirming processes. In many contexts, healthcare implicates networks beyond the individual patient. Consequently, healthcare practitioners must be able to enquire into the extent to which family and/or allies are engaged in care. Moreover, this is not a question that applies to a discrete time, but must be revised across patients' lifespans, development, changes their pathways of care (Byrne 2015; Feinberg 2001; Poteat 2013; Siverskog 2015).

In the context of indigenous gender engagement, in Aotearoa New Zealand and elsewhere, there are strategies that encompass indigenous trans peoples' perspectives, skills, values and language(s). These approaches offer information, guidance and processes to inform and challenge current international health practices, training, research and theory (Ashleigh 2014; Bith-Melanders et al. 2010; Byrne 2015; Kerekere. 2017; Nia Nia et al. 2017). Given the barriers that can be created by refusing to acknowledge and respect indigenous viewpoints, future work should address how healthcare providers can be facilitated to engage in these approaches (Ansara 2012; Winter et al. 2016). While these issues are by no means the extent of all aspects that should be carefully questioned in healthcare contexts, we believe that these are some key areas of concern that might form the basis of the continued evolution of trans and nonbinary healthcare praxis.

Developing Minoritarian Approaches

Much of trans people's healthcare is based around diagnostic manuals, psychometric measures, standards of care, and models and vocabularies of engagement, based on quantitative reductions of who trans people are, how they ought to be treated, and what the limitations of possible ways of existing are. This might be unproblematic for a disease model when symptoms are incontrovertibly harmful to their host and when the disease evolves slowly, but trans people's gender is not pathological and their gender can rapidly evolve beyond the scope of any model; a rhizomatic multiplicity. The language of treatment seeks to name a condition, rather than talking through a process, which can be incompatible for different people in various ways. Even with the increased flexibility that has been integrated into models, manuals and measures are designed to guide providers around who qualifies as trans, whether they may receive the treatment they desire, and what they are expected to develop into. This is problematic in many ways, including the assumption that trans people necessarily develop into their gender, rather than engendering congruency with what they already are. Flexible parameters are still parameters and the central idea we are trying to articulate is that, no matter how understandable it is that some practitioners would like a model to provide some assurance that they are providing "best practice" healthcare, what is "best" varies between clients and over time.

Until recently, trans healthcare models have pathologized trans people for existing (Burke 2011), rather than addressing trans people as part of natural variation. Moreover, the methods with which these models are created are based on assumptions that gender and transitional processes are reducible and generalizable, wherein any aberrant characteristic is deemed an outlier, and the diverse populations that they measure are typically represented by an average (Ghasemi and Zahediasl 2012). The heavy reliance on quantitative methods to build models of assessment and interaction is part of the production of normativities, wherein individuals are measured against others and averages, rather than encouraging engagement with individuals on their own terms. It is beyond the scope of this chapter to debate the contextual utility of quantitative approaches (which are more broadly and effectively discussed by Hwahng 2016), but we hope to encourage the reader to recognize that the models that quantitative methods produce are not representative of individuals, meaning they have a limited ability to inform individualized, client-centered practice.

Transgender, trans, trans★, transsexual, third gender, bigender, non-binary, genderqueer, and agender (to limit our commentary to a few examples of Anglo-centric Western labels) are approximations, like the statistics that attempt to summarize them. These labels are real insofar as they are meaningful to the people who are using them and assuming them as an identity produces material and discursive differences in those individuals' lives. In practical terms, this means that, as healthcare providers, engaging with an individual's gender requires that we simultaneously acknowledge the normativities that people may or may not operate with/against, while at the same time withholding assumptions around how or whether individuals are engaging with them. All people are minoritarians, grappling with gender normativities. A healthcare provider's position is to facilitate their patients' desired existences, insofar as gender affirmation procedures are concerned, address any issues and ensure that individuals are seen for who they are, in the ways they need to be seen (including everything that contextualizes their gender), rather than on the terms of and according to the extent that a model can accommodate who they are. This is the corollary of a system that is based on informed consent to allow people to embody their gender in the ways that are congruent with them. These are significant decisions and genuine informed consent requires this level of investment in mutual practitioner-client understandings. Practitioners should be most concerned about situations patients feel they were misinformed about the full scope of possibilities, rather than worried about patients regretting transition in the cases where the process did not meet their expectations in spite of informed consent (Ashley 2019).[12] It is the responsibility of healthcare providers to be capable of providing space for conversations with their clientele around how they can engage with transnormativities when this might be something that benefits the client.

These conversations cannot occur without the recognition that, insofar as imperial or (post)colonial contexts are concerned, our dominant institutions emerged largely, if not solely, from systems that prioritized white, cisheteronormative, able-bodied, wealthy, privileged ways of being. In Aotearoa New Zealand – a nation with a colonial history and an indigenous population, similar to many places in the world – Māori are beginning to (re)engage with a range of identities (takatāpui, whakawahine, hinehi, hinehua, tāhine and tangata ira tane; Kerekere 2017; To Be Who I Am 2008). We witness similar accounts from indigenous people elsewhere (binaohan 2014; Boellstorff et al. 2014; Roen 2006). Decolonized means of engaging with gender are essential for genuine and effective treatment, if we are to hold these institutions to prioritizing the well-being of individuals, rather than merely fitting them into meanings that are intelligible to dominant societal systems (binaohan 2014; Roen 2006). This means that healthcare providers need to be prepared to gain insight into contextual knowledge that may inform their clients' experiences and identities; religion, education, class, (dis)ability, sexuality, gender and whatever else is relevant to patients. Rather than treating these identity categories as stable indicators of how to engage with trans people, they are points of departure from which one can begin to understand how to better accommodate trans people through conversations that do not assume white, cisheteronormative, wealthy, educated, ableist standards. To achieve this, one needs to first contextualize how gender has been conceptualized, previously and presently. To move beyond the limitations of these paradigms, one needs to (re)conceive what it means to experience gender.

Despite its inability to capture the variation in human experience and embodiment of gender, the binary system has provided an effective mechanism of social constraint, wherein dissent is severely policed and punished. Linstead and Pullen (2006) highlight two ways in which the binary system of gender has attempted to be reterritorialized: Multiplicities of the Same and Multiplicities of the Third. The Multiplicities of the Same reflect the spirit of movements that have sought to expand what it means to be male or female. This is characteristic of feminist movements that have sought to generate influence through unity under the banner of womanhood, but in doing so, have reified and reinforced the gender binary that perpetuated this hegemony in the first place. To suggest that one can summarize almost half of the world's population under one label is inadequate when it comes to responding to people's multiplicitous needs.

The Multiplicities of the Third are characterized by the normalization of new categories (Linstead and Pullen 2006), as we have observed in homo- and trans-normativities. Unintelligible experiences are made intelligible through an appeal to existing norms, generating hegemony determined by the degree to which individuals may achieve these new norms. We see this in the slow expansion of the gender, sex and sexual initialism: LGBTQIA+. . . .

It is not that these people did not exist before, but that they are being (re) defined (and therein reduced to an initial) through normative vocabularies. Puar (2015) discusses the way that this neoliberal vocabulary often reduces trans people down to their utility or ability to perform normative roles; to be "pieced", as Puar describes it. The "T" in LGBTQIA+ becomes a brand that is sold, paired with "opportunities" that mask constraint, and "success" is defined by the degree of conformity. Because transnormativities are entangled with cisnormativity (as well as classism, ableism, white supremacy, and other normative structures), like the Multiplicities of the Same, this diversification of identity labels is condemned to reinforce the dominance of the gender binary.

To move beyond majoritarian gender binaries, we need a mechanism of engaging with gender from a minoritarian perspective – through the relations that produce the individual,[13] rather than the normativity. We took inspiration from Linstead and Pullen (2006), who employed Deleuze and Guattari's (1987) concept of rhizome to (re)theorize gender. Linstead and Pullen (2006: 6) describe gender as "a constant journey with no destination"; a middle with no beginning or end. Therefore, engaging with people with regard to their gender means meeting them at their middle, on their terms. As individuals' bodies move through the world, they are influenced by other bodies, objects, ideas, events and absences. As well as being (re)constituted by their worlds, they (re)constitute the worlds around them. In the context of healthcare, this is made up of all the elements that comprise individuals' experiences leading up to, during, and after gender affirming treatment (or the lack thereof), accounting for all of the different ways these aspects of their lives are assembled. These encounters allow for the multiplicity of possible of ways of being in the world to proliferate; nuanced engagements generating new possibilities. Embracing rhizomatic thinking requires recognizing that everything that influences a person's gender is defined in relation to everything else that makes up who they are, so one cannot simply isolate one aspect.

Subsequently, Linstead and Pullen (2006) recommend that we treat rhizomes (and, for our purposes, genders) as cartographies. When seeking to meet the needs of (trans) people, this should be employed in a way that recognizes that the topographies or composition of an individual's gender may shift over time and context. They are not essentializable, generalizable or paradigmatic. As a health professional and their client's relationship develops, it is necessary to continue mapping the client's gender(s) together, accounting for ongoing power dynamics, to make sure the issues addressed are relevant to the patient. This requires critical engagement with the modes and models that health professionals have been educated through, to continually reassess whether these majoritarian arrangements are relevant or useful to their patients. Some health professionals are already engaging in these sorts of practices, but this chapter seeks to provide the health profession, as a whole, with the tools to disengage from normative modes of treatment that do not serve patients outside these norms. By engaging each individual's gender as a process of becoming through which the practitioner becomes part of by virtue of their relation to one another, healthcare professionals may move beyond (re)enforcing transnormativities through their practices.

Notes

1 We employ the term "non-conforming" over "diverse" in opposition to the idea that only non-cisgender people can be diverse. Also, this term gels better with notions of normativity.
2 Notably, we avoid the term "best" or "good" practice because we do not believe an absolute standard of healthcare should be the goal of practitioners whose patients' definition of "best" or "good" is inevitably varied and fluid.
3 Cisgender means to identify with the gender one is assigned at birth.
4 Māori are the indigenous people of Aotearoa New Zealand; Pākehā refers to New Zealanders of white European extraction.
5 "Passing "typically refers to trans people being perceived as a cisgender male or female as a result of their performance of gender roles. However, in the context of non-binary people, passing may be a matter of being perceived as androgynous.
6 Notably, this distinction assumes a gender binary.
7 "Respectability politics" refers to disenfranchised individuals performing the norms of the dominant group (e.g., whiteness) in the hope of being perceived as "respectful" and therefore receiving better treatment (e.g., reduced institutionalized violence). Valorization of dominant norms inherently Others members of the disenfranchised group who cannot or will not perform dominant norms, thereby affording conformists greater access to power at the expense of the Othered. Notably, this approach often undermines minoritarian power in the long term because it reinforces the dynamics that created the inequality in the first place.
8 Again, we are not critical of these things in themselves, but are conscious of the effects on those who are excluded as a consequence.
9 While beyond the scope of this chapter, this does present some challenging questions around the possibilities of informed consent in these terms.
10 Note that not all non-binary people identify as trans.
11 This refers to a critique of the medical ethic of "do no harm", which is often used to argue against treatment because of risks. In this case, the argument is that doing nothing can cause harm to trans people denied healthcare. This line of reasoning is recognised in the Guidelines for Gender Affirming Healthcare (Oliphant et al. 2018).
12 While these circumstances are not ideal and may be detrimental to an individual's well-being, they are the risks that any patient would consent to for any procedure.
13 "Individual experience" does not seek to privilege Western individualism, recognizing that individual experience is situated within the context of community, class, culture, (dis)ability, and so on. What we advocate in the provider-patient dyad is that interactions are based on an individual's reality, rather than stereotype or normativity.

References

Aizura, A. (2016). Affective vulnerability and transgender exceptionalism: Norma ureiro in transgression. In *Trans studies: The challenge to hetero/homo normativities*. Edited by Martinez-San Miguel, Y., & Tobias, S. Rutgers University Press. (8), 122–137.

Altemeyer, B. (1981). *Right-wing authoritarianism*. University of Manitoba Press, Winnipeg.

American Psychological Association (2010). Ethical principles of psychologists and code of conduct (Including 2010 and 2016 Amendments). Retrieved 21 November 2021 from https://www.apa.org/ethics/code.

American Psychological Association (2015). Guidelines for psychological practice with transgender and gender nonconforming people. *American Psychologist, 70*(9), 832–864.

Ansara, Y. G. (2012). Cisgenderism in medical settings: Challenging structural violence through collaborative partnerships. In I. Rivers & R. Ward (Eds.), *Out of the ordinary: Representations of LGBT lives* (pp. 93–112). Cambridge: Cambridge Scholars Publishing.

Ashleigh, F. (2014). *La e ola malamalama I lou fa'asinomaga: A comparative study of the fa'afafine of Samoa and the whakawahine of Aotearoa/New Zealand*. Victoria University of Wellington, New Zealand.

Ashley, F. (2019). Gatekeeping hormone replacement therapy for transgender patients is dehumanising. *Journal of Medical Ethics, 45*(7), 480–482.

Bilodeau, B. (2005). Beyond the gender binary: A case study of two transgender students at a Midwestern research university. *Journal of Gay & Lesbian Issues in Education, 3*(1), 29–44.

binaohan, b. (2014). *Decolonizing trans/gender 101*. Biyuti Publishing, Toronto.

Bith-Melanders, P., Sheoran, B., Sheth, L., Bermudez, C. Drone, J., Wood, W., & Schroeder, K. (2010). Understanding sociocultural and psychological factors affecting transgender people of colour in San Francisco. *Journal of the Association of Nurses in AIDs Care, 21*(3), 207–220

Boellstorff, T., Cabral, M., Cárdenas, M., Cotten, T., Stanley, E. A., Young, K. and Aizura, A. Z. (2014). Decolonizing transgender: A roundtable discussion. *Transgender Studies Quarterly, 1*(3), 419–439.

Brown, G. (2009). Thinking beyond homonormativity: Performative explorations of diverse gay economies. *Environment and Planning A, 41*(6), 1496–1510.

Brown, G. (2012). Homonormativity: A metropolitan concept that denigrates "ordinary" gay lives. *Journal of Homosexuality, 59*(7), 1065–1072.

Budge, S. L. (2015). Psychotherapists as gatekeepers: An evidence-based case study highlighting the role and process of letter writing for transgender clients. *Psychotherapy, 52*(3), 287.

Butler, J. (1990). *Gender trouble: Feminism and the subversion of identity*. New York: Routledge,.

Burke, M. C. (2011). Resisting pathology: GID and the contested terrain of diagnosis in the transgender rights movement. In *Sociology of diagnosis* (pp. 183–210). Emerald Group Publishing Limited: Bingley, UK.

Byrne, J. (2015). Blueprint for the provision of comprehensive care for trans people and trans communities in Asia and the Pacific, 96–101

Callen Lorde, Community Health Centre (2016). Protocols for the provision of hormone therapy, 1–38.

Catalano, D. C. J. (2015). "Trans enough?" The pressures trans men negotiate in higher education. *TSQ: Transgender Studies Quarterly, 2*(3), 411–430.

Charles, K. K., & Hurst, E. (2003). The correlation of wealth across generations. *Journal of political Economy, 111*(6), 1155–1182.

Chisolm-Straker, M., Jardine, L., Bennouna, C., Morency-Brassard, N., Coy, L., Egmba, O. M., & Shearer, P. (2017). Transgender and gender nonconforming in emergency departments: A qualitative report of patient experiences. *Transgender Health, 2*(1), 8–16

Clark, T. C., Lucassen, M. F. G., Bullen, P., Denny, S. J., Fleming, T. M., Robinson, E. M., & Rossen, F. V. (2012). The health and well-being of transgender high school students: Results from the New Zealand adolescent health survey (youth'12). *Journal of Adolescent Health, 55*, 93–99.

Cruz, T. M. (2014). Assessing access to care for transgender and gender nonconforming people: A consideration of diversity in combating discrimination. *Social Science & Medicine, 110*, 65–73

Cundill, P.Dr., & Wiggins, J. (2017). Protocols for the initiation of hormone therapy for trans and gender diverse patients. Equinox Gender Diverse Health Centre. Australia

Deleuze, G. (1992). Postscript on the societies of control. October, 59, pp. 3–7.

Deleuze, G., & Guattari, F. (1987). *A thousand plateaus*. Trans. Brian Massumi. Minneapolis: Te University of Minnesota Press.

Deutsch, M. D. (2016). *Guidelines for the primary and gender-affirming care of transgender and gender nonbinary people*. Centre of Excellence for Transgender Health, Department of Family and Community Medicine, University of California, San Francisco, 2nd edition.

Dowling, A. (2017). Meet TV's first non-binary-gender character: Asia Kate Dillon of showtime's 'billions'. The Hollywood Reporter. Retrieved 19 July 2017, from www.hollywoodreporter.com/live-feed/meet-tvs-first-binary-gender-character-asia-kate-dillon-showtimes-billions-979523

Drescher, J., Cohen-Kettenis, P., & Winter, S. (2012). Minding the body: Situating gender identity diagnoses in the ICD-11. *International Review of Psychiatry*, 24(6), 568–577.

DSM-V (2013). American psychiatric association: Diagnostic and statistical manual of mental disorders (5th ed.). Arlington: American Psychiatric Association.

Duggan, L. (2002). The new homonormativity: The sexual politics of neoliberalism. *Materializing Democracy: Toward a Revitalized Cultural Politics*, 175–194.

Feinberg, L. (2001). Trans health crisis: For us it's death. *American Journal of Public Health*, 9(0), 897–900

Finch, S. D. (2015). 8 things non-binary people need to know. (2015). Lets queer things up! Retrieved 19 July 2017, from https://letsqueerthingsup.com/2015/03/15/8-things-non-binary-people-need-to-know/

Foucault, M. (1977). *Discipline and punish*, trans. Alan Sheridan (New York: Vintage, 1979), 242.

GATE, Global Action for Trans Equality. (2013). Critiaue and alternative proposal to the 'gender incongruence of childhood' category in ICD-11. GATE Civil Society Expert Working Group, Buenos Aires, April 4–6, 2013.

Gender Minorities Aotearoa. (2015). Available online : https://genderminorities.com/find-transgender-info-services/resources/human-rights-and-discrimination/

Ghasemi, A., & Zahediasl, S. (2012). Normality tests for statistical analysis: A guide for non-statisticians. *International Journal of Endocrinology and Metabolism*, 10(2), 486.

Glick, P., & Fiske, S. T. (1996). The ambivalent sexism inventory: Differentiating hostile and benevolent sexism. *Journal of Personality and Social Psychology*, 70(3), 491.

Glick, P., & Fiske, S. T. (1997). Hostile and benevolent sexism: Measuring ambivalent sexist attitudes toward women. *Psychology of Women Quarterly*, 21(1), 119–135.

Glick, P., Lameiras, M., Fiske, S. T., Eckes, T., Masser, B., Volpato, C.,. & Castro, Y. R. (2004). Bad but bold: Ambivalent attitudes toward men predict gender inequality in 16 nations. *Journal of Personality and Social Psychology*, 86(5), 713.

Halberstam, J. (2005). *In a queer time and place: Transgender bodies subcultural lives*. New York University Press: New York, USA.

Hughto, J. M. W., Reisner, S. L., & Pachankis, J. E., 2015. Transgender stigma and health: A critical review of stigma determinants, mechanisms, and interventions. *Social Science & Medicine*, 147, 222–231.

Hwahng, S. J. (2016). Adventures in trans biopolitics: A comparison between public health and critical academic research praxis. In Martinez-San Miguel, Y., & Tobias, S. (Eds.) *Trans studies: The challenge to hetero/Homo Normativities* (pp. 191–214). New Brunswick, NJ and London: Rutgers University Press.

Hyde, Z., Doherty, M., Tilley, P. J. M., McCaul, K. A., Rooney, R., & Jancey, J. (2014). *The first Australian national trans mental health study: Summary of results*. Perth, Australia: School of Public Health, Curtin University.

Irving, D. (2012). Elusive subjects: Notes on the relationship between critical political economy and trans studies. In Anne Enke (Ed.), *Transfeminist perspectives in and beyond transgender and gender studies* (pp. 153–169). Temple University Press: Philadelphia, USA.

Jones, T., Smith, E., Ward, R., Dixon, J., Hillier, L., & Mitchell, A. (2016). School experiences of transgender and gender diverse students in Australia. *Sex Education*, 16(2), 156–171

Joynt, C., & Bryson, M. K. (2012). RESISTERECTOMY by Chase Joynt. (2017). YouTube. Retrieved 19 July 2017, from www.youtube.com/watch?v=nPLdJMm0TPA

Kerekere, E. (2017). Takatāpui | A resource hub. Retrieved 19 July 2017, from www.takatapui.com/#home

Kunzel, R. (2014). The flourishing of transgender studies. *TSQ: Transgender Studies Quarterly*, 1(1–2), 285–297.

Linstead, S., & Pullen, A. (2006). Gender as multiplicity: Desire, displacement, difference and dispersion. *Human Relations*, 59(9), 1287–1310.

Lubbers, E. (2015). Transgender representation and Japan: Manga's potential for disrupting gender (Master's thesis).

Lykens, J. E. (2016). The transgender binary and genderqueer health (Doctoral dissertation, San Francisco State University).

Martinez-San Miguel, Y., & Tobias, S. (2016). Introduction: Thinking hetro/homo normativities. *Trans Studies: The Challenge to Hetero/Homo Normativities*, 1

Matte, N. (2014). *Historicizing liberal american transnormativities: Medicine, media, activism, 1960–1990*. Canada: University of Toronto.

Meadow, T. (2017). Whose chosenness counts? The always-already racialized discourse of trans – response to Rogers Brubaker. *Ethnic and Racial Studies*, 40(8), 1306–1311.

Meyer, I. H. (2015). Resilience in the study of minority stress and health of sexual and gender minorities. *Psychology of Sexual Orientation and Gender Diversity*, 2(3), 209.

Nia Nia, W., Bush, A., & Epston, D. (2017). *Collaborative and indigenous mental health therapy: Tātaihono – stories of Māori healing and psychiatry* (pp. 21–37). New York and London: Routledge Taylor and Francis Group.

Noble, B. J. (2012). Our bodies are not ourselves: Tranny guys and the racialized class politics of incoherence. *Queerly Canadian: An Introductory Reader in Sexuality Studies*, 35.

Oliphant, J., Veale, J., Macdonald, J., Carroll, R., Johnson, R., Harte, M., Stephenson, C., & Bullock, J. (2018). *Guidelines for gender affirming healthcare for gender diverse and transgender children, young people and adults in Aotearoa New Zealand*. Transgender Health Research Lab, University of Waikato: Hamilton.

Pega, F., & Veale, J. (2015). The case for the world health organization's commission on social determinants of health to address gender identity. *American Journal of Public Health*, 105(3), 58–62.

Perez, H. (2005). You can have my brown body and eat it, too! *Social Text*, 84, 171.

Pitts, M., Couch, M., Croy, S., Mitchell, A., & Mulcare, H. (2009). Health service use and experiences of transgender people: Australian and New Zealand perspectives. *Gay and Lesbian Issues and Psychology Review*, 5(3), 167.

Poteat, T., German, D., & Kerrigan, D. (2013). Managing uncertainty: A grounded theory of stigma in transgender health care encounters. *Social Science and Medicine*, 884, 22–29.

Puar, J. K. (2007). *Terrorist assemblages: Homonationalism in queer times*. Duke University Press: Durham, USA.

Puar, J. K. (2013). Rethinking homonationalism. *International Journal of Middle East Studies*, 45(2), 336–339.

Puar, J. K. (2015). Bodies with new organs becoming trans, becoming disabled. *Social Text*, 33(3 124), 45–73.

Purdie-Vaughns, V., & Eibach, R. P. (2008). Intersectional invisibility: The distinctive advantages and disadvantages of multiple subordinate-group identities. *Sex Roles*, 59(5), 377–391.

Reed, T. (2016). GIRES: e-learning for transgender health training. *The Lancet*, 388(10042), 333.

Reisner, S., Poteat, T., Keatley, J., Cabral, M., Mothopeng, T., Dunham, E., Holland, C. E., Max, R., & Baral, S. D. (2016). Global health burden and needs of transgender populations: A review. *The Lancet*, (388), 412–436.

Richards, C., Bouman, W. P., Seal, L., Barker, M. J., Nieder, T. O., & T'Sjoen, G. (2016). Non-binary or genderqueer genders. *International Review of Psychiatry*, 28(1), 95–102.

Roberts, T. K., & Fantz, C. (2014). Barriers to quality health care for transgender population. *Clinical Biochemistry*, (47), 983–987

Roen, K. (2006). Transgender theory and embodiment: The risk of racial marginalisation. *Journal of Gender Studies*, 10(3), 253–263.

Santos, A. C. (2013). Are we there yet? Queer sexual encounters, legal recognition and homonormativity. *Journal of Gender Studies*, 22(1), 54–64.

Sekuler, T. (2013). Convivial relations between gender non-conformity and the French nation-state. *L'Esprit Créateur*, 53(1), 15–30.

Sibley, C. G., Overall, N. C., & Duckitt, J. (2007). When women become more hostilely sexist toward their gender: The system-justifying effect of benevolent sexism. *Sex Roles*, 57(9–10), 743.

Silverman, M. (2008). Issues in access to healthcare by transgender individuals. *Women's Rights Law Reporter*, 30, 347.

Siverskog, A. (2015). Ageing bodies that matter: Age, gender and embodiment in older transgender people's life stories. *NORA-Nordic Journal of Feminist and Gender Research*, 23(1), 4–19.

Snelgrove, J. W., Jasudavisius, A. M., Rowe, B. W., Head, E. M., & Bauer, G. R. (2012). "Completely out-at-sea" with "two-gender medicine": A qualitative analysis of physician-side barriers to providing healthcare for transgender people. *BMC Health Services Research*, 12, 110.

Snorton, C. R., & Haritaworn, J. (2013). Trans necropolitics: A transnational reflection on violence, death, and the trans of color afterlife. *The Transgender Studies Reader*, 2, 66–76.

Statistics New Zealand (2018). Sex, gender, and sexual orientation. *Stats.govt.nz*. Retrieved 18 February 2018, from www.stats.govt.nz/reports/sex-gender-and-sexual-orientation.

Statistics New Zealand (2021). 2023 Census first to collect gender and sexual identity from everyone in Aotearoa New Zealand. Stats.govt.nz. Retrieved 08 December 2021, from https://www.stats.govt.nz/news/2023-census-first-to-collect-gender-and-sexual-identity-from-everyone-in-aotearoa-new-zealand

Stryker, S., Currah, P., & Moore, L. J. (2008). Introduction: Trans-, trans, or transgender? *Women's Studies Quarterly*, 11–22.

Taylor, Y., Hines, S., & Casey, M. (Eds.). (2010). *Theorizing intersectionality and sexuality*. Palgrave Macmillan: New York.

To Be Who I Am (2008). New Zealand Human Rights Commission. Retrieved 20 July 2017, from www.hrc.co.nz/files/5714/2378/7661/15-Jan-2008_14-56-48_HRC_Transgender_FINAL.pdf

Vance, S. R., Halpern-Felsher, B. L., & Rosenthal, S. M. (2015). Health care providers' comfort with and barriers to care of transgender youth. *Journal of Adolescent Health*, (56), 251–253.

Veale, J. F., Watson, R. J., Peter, T., & Saewyc, E. M. (2016). Mental health disparities among Canadian transgender youth. *Journal of Adolescent Health*, 60(1), 44–49.

Veale, J. F., Byrne, J., Tan, K. K. H., Guy, S., Yee, A., Nopera, T. M.-L., & Bentham, R. (2019). *Counting Ourselves: The health and wellbeing of trans and non-binary people in Aotearoa New Zealand*. Transgender Health Research Lab University of Waikato: Hamilton, New Zealand.

Vincent, B. (2019). Breaking down barriers and binaries in trans healthcare: The validation of non-binary people. *International Journal of Transgenderism*, 20(2-3), 132-137.

Weiss, M. (2011). *Techniques of pleasure: BDSM and the circuits of sexuality*. Duke University Press: Durham, USA.

White Hughto, J. M., Reisner, S. L., & Pachankis, J. E. (2015). Transgender stigma and health: A critical review of stigma determinants, mechanisms and interventions. *Social Science & Medicine*, 147, 222–231.

Winter, K. (2008). Gender madness in American psychiatry: Essays from the struggle for dignity. GID Reform Advocates, Dillion, CO, 101–117.

Winter, S., Diamond, M., Green, J., Darasic, D., Reed, T., Whittle, S., & Wylie, K. (2016). Transgender people: Health at the margins of society. *The Lancet*, 388(10042), 390–400.

WPATH (2021). Standards of Care Version 8. Retrieved 21 November 2021, from: https://www.wpath.org/soc8.

WPATH SOC v.7 (2001). The Harry Benjamin International Gender Dysphoria Association's standards of care for gender identity disorders, 6th version; 2001. Retrieved 31 October 2016, from www.cpath.ca/wp-content/uploads/2009/12/WPATHsocv6.pdf.

WPATHSOC v.6 (2011). Standards of care for the health of transsexual, transgender, and gender nonconforming people. The World Professional Association for Transgender Health. Retrieved 21 December 2021, from Retrieved 31 October 2016, from.

Wylie, K., Knudson, G., Kahn, S. I., Bonierbale, M., Watanyusakul, S., & Baral, S. (2016). Serving transgender people: Clinical care considerations and service delivery models in transgender health. *The Lancet*, 388(10042), 401–411.

Yogakarta Principles. (2006). Retrieved 17 January 2018, from http://yogyakartaprinciples.org/.

Yogakarta Principles Plus 10. (2017). Retrieved 17 January 2018, from http://yogyakartaprinciples.org/.

9 Cisnormative and Transnormative Misgendering

Holding Gender Minorities Accountable to Gender Expectations in Mainstream and Trans/Queer Spaces

Sonny Nordmarken

Policy changes affirming transgender identities in some women's colleges and the Girl Scouts and Boy Scouts of America point to increasing institutional integration of trans people in the United States, a weakening of hegemonic gender ideology, and the emergence and affirmation of an alternative gender paradigm. However, anti-transgender "bathroom bills" have passed in several states, reflecting an increase in what Viviane Namaste called "institutional erasure" (2000). It is, then, an ideal time to examine the conditions under which gender minorities continue to experience misrecognition in social interactions.

Gender minorities typically use the term "misgendering" to refer to their experiences being read as the gender they were assigned at birth, rather than the gender they identify as. Misgendering actions can be understood as "microaggressions" (Nordmarken 2014; Nordmarken and Kelly 2014) or "discursive aggressions" (shuster 2017) that hold gender minorities accountable to gender expectations. The term "misgendering" frames individuals' identity claims as correct by framing attributors' readings as incorrect. The term also challenges dominant gender attribution norms, which people use in attempts to hold gender minorities accountable to the dominant gender paradigm. In this chapter, I use the term "misgendering" to describe all experiences of gender misreading, in order to bring gender minorities' perspectives to bear in advancing scholarly conversations. As I illustrate, misgendering actions hold gender minorities accountable to expectations of both dominant, *cisnormative, heteronormative ("cishet")* and counter-hegemonic, *transnormative gender paradigms*. This chapter draws on research that examined 21 gender minorities' accounts of misgendering within mainstream, or cishet-dominated, and trans/queer public spaces. I consider these individuals' reflected appraisals, or perceptions of others' perceptions of them, to unpack the factors they perceived contributed to their misgendering. This research reveals several complex ways that cisnormative and transnormative gender frames operated in these social spaces, limiting recognition. I find that gender minorities encountered cisnormative misgendering in cishet-dominated spaces, where perceivers focused on their outward bodily appearance and record of gender assignment, while in trans/queer spaces, their

DOI: 10.4324/9781315613703-9

aesthetic appearance, body shape, pronouns and race appeared to contribute to transnormative misgendering.

Cisnormative and Trans Gender Paradigms

Gender has been theorized as a primary cultural frame, which organizes social relations (Ridgeway 2009). Gender beliefs based in this frame shape inter-actional behavior. Some gender beliefs cast men, for example, as dangerous perpetrators of violence and women as vulnerable to victimization (Hollander 2001). Ridgeway conceptualized gender as a single, overarching frame, but as there are multiple gender frames and paradigms, which I will discuss, I have distinguished the frame discussed by Ridgeway and others as the "dominant gender paradigm" in the contemporary United States (Nordmarken 2019). Beliefs that gender is signified by genitals and is a "natural matter of fact" (Gar-finkel 1967: 125) are defining assumptions of the dominant gender paradigm. Some related assumptions are that gender and gendered features are immutable (Medin and Ortony 1989), and therefore, that gender identities always align with birth gender assignments. These constructions work to cast culturally conforming gender identities, bodies and expressions as normal, natural, real and of value, figuring individuals who identify with their assigned gender as superior to and more authentic than those who do not (Serano 2007; Kelly 2012). These notions deny the existence of gender minorities, figuring them as "pretenders" and "deceivers" (Bettcher 2007). Akin to heteronormativity and heterosexism, scholars have termed this dominant conceptual formation "cisnormativity" (Kelly 2012) and its social manifestation "cissexism" (Serano 2007). The dominant gender frame is a cisnormative gender frame.

Trans communities contest the cisnormative paradigm by distinguishing internal gender identity from the material body, granting legitimacy to such self-ascribed gender identities over assigned or attributed gender (Bornstein 1994; Nordmarken 2019). Prosser showed how transsexual autobiographers understood the body as a kind of "second skin" encasing, concealing and mis-representing one's self-image, which is constructed as the real, "inner body" or gender identity within (1998). I have used the term "binary trans paradigm" to describe this prominent, cultural, binary-conceptual model of trans identity (Nordmarken 2019). Going beyond the "wrong-body" metaphor, what I have called the "queer trans paradigm" (Nordmarken 2019) honors a plurality of gender identities, emphasizing that identity, body and appearance can be non-binary and/or fluid (e.g., Bornstein 1994). Segments of trans communities that understand gender in this way conceptually disaggregate body, pronouns and gender presentation from gender identity, which means that a trans individual may make no changes to their appearance (Nordmarken 2019). The logic of an identity within (Prosser 1998) applies to both trans paradigms, positioning indi-viduals as epistemic authorities about their own gender identities. Both trans paradigms challenge pervasive cisnormative cultural assumptions that gender is definitively aligned with "sex" and that an audience's perception of a person is more legitimate than that individual's own self-identification (Serano 2007).

As frames shape behavior (Ridgeway 2009), and attributing gender is an interactional practice, assumptions characterizing the dominant gender paradigm – that gender is fixed, signified by genitals, and a "natural matter of fact" (Garfinkel 1967: 125), and that people identify with the gender they were assigned at birth – shape how gender is attributed. In gender's "routine, methodical and recurring accomplishment" (West and Zimmerman 1987: 126), where social actors interactionally, outwardly display cues and read others' cues to attribute gender (West and Zimmerman 1987), attributors working within the dominant gender paradigm consider clothing, hairstyle, mannerisms, genitals (Kessler and McKenna 1978), tone of voice (Goffman 1977) and bodily characteristics (Dozier 2005) important gender cues. These outward gender cues, in the dominant gender paradigm, are taken as biological truths/referents (i.e., one's sex). Thus, the dominant gender paradigm shapes the cultural practice of interpreting a person's appearance to attribute a gender to them, and actors hold each other accountable to these expectations.

Operating in this context, gender minorities have used various strategies to manage others' impressions of them, modifying their presentation of self in order to be appropriately understood. Studies have shown how gender minorities present themselves through adornment and bodily comportment (Garfinkel 1967, West and Zimmerman 1987) and alter their bodies (Schilt 2010; Schilt and Windsor 2014). Certain segments of trans communities have developed social practices to facilitate recognition regardless of visual gender presentations: they verbally identify their own third person gender pronouns (Zimman 2017; Nordmarken 2019). These verbal practices operationalize trans paradigms and enable interaction partners to honor an array of gender identities, non-identities and pronouns, as many individuals do not signal their identities through adornment (Nordmarken 2019).

Although the dominant gender paradigm continues to structure the majority of institutions, certain aspects of transgender paradigms have been incorporated into some traditionally cishet-dominated organizations, such as schools and youth organizations (Carapezza 2014; Eversley and Koloff 2017). Trans paradigms have also begun to be incorporated into some cishet-dominated interactional spaces. Family members (Meadow 2011; Norwood 2013; Rahilly 2015), partners (Pfeffer 2014; Tompkins 2014), college instructors (Wentling 2015), male inmates (Jenness and Fenstermaker 2014) and news consumers (Westbrook and Schilt 2014) have honored trans identity claims and pronouns.

Analyzing autobiographies, Stone (1991) inaugurated the contemporary field of transgender studies by arguing that trans people's formulations of the "wrong body" discourse are ineffective in trans struggles for desubjugation, as they maintain pathologizing interpretations of trans people and the idea that gender is binary, and thus fail to adequately challenge the cisnormative meaning system. Along the same lines, Johnson (2016) used the term "transnormativity" to describe the interpretive framework where trans people's legitimacy is dependent on medical standards, which marginalizes those who do not desire medical transition. Taking a different approach, Aizura (2006) and Snorton and Haritaworn (2013) used the term "transnormativity" akin to "homonormativity" (Duggan 2003), to describe a broadly assimilationist politics, where trans individuals seek inclusion in state institutions complicit in capitalist and imperialist projects. In this chapter, I apply the term "transnormativity" to the production of meaning in social interactions, to consider how, regardless of intention, gender attributions that reflect narrow understandings or misunderstandings of varied instantiations of transness can act as regulatory practices. In this research, I ask, "what ideological notions are gender minorities held accountable to in their experiences of misgendering in cishet-dominated and trans/queer spaces"?

Methods

This chapter draws on semi-structured interviews with 21 gender minority individuals ranging from 40 minutes to three hours in the San Francisco Bay Area in 2011, 2012 and 2017. A sample of this size is limited and the findings should be generalized neither to all gender minorities nor to their various communities. I chose San Francisco as the field site in order to maximize access to participants who had regular interactional experiences in both trans/queer spaces and cishet-dominated spaces. San Francisco is home to a sizable gender minority population where frequent, public events are organized and attended by majority trans and queer community members.

To account for various kinds of misgendering experiences, I interviewed a gender-diverse group, including binary and non-binary identified individuals and both those who were either visibly gender atypical or not. Five individuals claimed identity labels classifiable as transmasculine, five claimed labels classifiable as transfeminine, six claimed various labels I classify as unconventional (e.g., drag performers, cross-dressers, genderqueer, gender nonconforming, gender variant, gender fluid, non-binary), four claimed both transmasculine and unconventional labels, and one claimed both transfeminine and unconventional labels. They were diverse in socioeconomic class and age, with six in their 20s, three in their 30s, seven in their 40s, two in their 50s and three in their early 60s. Informants were particularly educated: seven held graduate degrees, four held bachelor's degrees and the remainder (10) had completed some college coursework. A majority (13) were white, and of the participants of color, two were African American, two were Asian American, one was Native American, one was Latinx and two claimed multiple racial identities – one was Asian American and Native American and one was Asian American and white. I recruited eight interviewees through personal networks and 13 through social media, outreaching to ethnicity- and heritage-based LGBTQ organizations in order to increase the racial and ethnic diversity of the sample.

In interviews, participants discussed their identities, everyday interactions with strangers and acquaintances in different contexts, instances where conflict over gender presentation or identity arose, and what happened in situations where others misgendered them. I transcribed the audio recordings and open- and selectively coded the transcripts for themes. All testimonials were included in the analysis. I analyzed participants' reflected appraisals (how they thought others saw them) and their descriptions of the perceptions others expressed when they encountered them.

As a trans person in the cis-dominated world, and a trans researcher in cis-dominated academe, I am an "outsider within" (Hill Collins 1986). Like Betsy Lucal, I have been both "socially male" and "socially female": I have been per- ceived and treated as male and female (1999). I have also been socially illegible, and I myself have experienced misgendering. As my positionality influences the knowledge I create (Mullings 1999), my particular history and context shape my research, data and interpretation (Angrosino 2005). My familiar- ity with trans communities may have enhanced my understanding of gender minorities and gendered processes and may have limited my analysis. It was important for me to, as Maykut and Morehouse (1994) suggest, acutely tune in to the meanings that others made out of their experiences while also noticing how my perspective influenced my perceptions. I attempted, as Rose (1985) advises, to be aware of my biases, so I could see how they shaped the data.

My identity also affected my findings as it shaped interactions with par- ticipants. As a trans person researching with trans people, I appear to be an insider – a "complete member researcher" (Adler and Adler 1987). My gender minority status likely enhanced my access to participants and the comfort it may have given them likely produced "richer" data (Talbot 1998–99). However, though I share some things with my participants, due to our differences (race, age, ability, gender), I will always be outsider in some ways to their identities and experiences. Although one would expect that interviewees who had less in common with me would be less inclined to share their stories, the length of interviews and emotional tenor of interactions between my participants and me did not appear to vary based on our differences. I also did not find any patterns in the richness of data across interviews, as most participants appeared to share their experiences openly. However, my identity as a white researcher likely shaped the racial make-up of the sample. Although I attempted to increase the racial diversity of the sample by outreaching intentionally to organizations for people of color, some recruitment strategies made the sample a conveni- ence sample, as I recruited in part using my own social networks, which were majority white. In addition, due to patterns of white researchers appropriating the stories of people of color, to the benefit of the researcher and many times causing harm to participants, trans people of color who heard about the study might have been reluctant to participate in the study.

Findings

Despite policy gains and increasing social awareness of gender minorities and trans paradigms, participants experienced others misgendering them in interactions in both mainstream and trans/queer spaces, even when they communicated how they wanted to be interpreted by adorning themselves, stating their pronouns, modifying their bodies and/or legally changing gender. Cisnormative misgendering took place as usual in cishet-dominated spaces. Accounts of misgendering in trans/queer spaces revealed transnormative misgendering, reflecting white racial framing and assumptions of cisnormative and transnormative paradigms.

The Cisnormative Priorities of Outward Bodily Appearance and Gender Assignment Record

Participants found few opportunities to verbally communicate their pronouns to people in mainstream public spaces and perceivers often ignored their clothing and hairstyle (gender presentation), instead evaluating outward bodily appearance, leading to misgendering. Jim, who wore a "men's" hairstyle and clothes and identified himself as a Korean and Cherokee, pre-transition, transgender male in his mid-40s about to start testosterone, shared:

> Now that I've decided to transition, I'm very dismayed that I'm still not quite passing. And I know I will after I start taking testosterone, but I've been getting lot of guys in particular calling me "little lady" and "ma'am" and I always mumble under my breath "I don't look anything like a little lady."

Like Jim, Lance, a white, fat-identified, genderqueer, gender fluid, femme, boy/transguy in his early 20s, who also wore short hair and "men's" clothes and had not made bodily changes, such as hormones or surgery, had a similar experience:

> On the elevator with two guys, they were like, "Are you sure you want to go to the fourth floor? That's a men's-only floor." I was like, "Yes, I'm a guy." I was so frustrated because . . . I knew I was going to meet a lot of people that day, so I wore a tee-shirt that said "Trans Pride" with a sticker with my middle name and "he" pronouns. I couldn't have been more explicit about my gender. I was binding, putting myself in an uncomfortable position so 500 people would recognize I'm a guy or at least I'm not a woman.

Another participant, Saulo, recalled when his coworker misgendered a person he read as transfeminine:

> [My coworker] looked at the person, and was like, "That guy is so androgynous." I was like, "What makes you think that they're a guy?" She was like, "Well, I can see he's a guy." The person she was talking about I felt was clearly presenting as transfeminine.

These accounts suggest that, consistent with Dozier's (2005) findings, when "conflicting" signs of gender appeared (e.g., "female" bodily appearance and "masculine" gender presentation), perceivers weighed bodily criteria more heavily than gender presentation criteria. This suggests that though clothing, hairstyle and other aspects of gender presentation often function as "cultural genitals" (Kessler and McKenna 1978), outward bodily appearance functioned in these cases more primarily as "cultural genitals". These experiences reflect the cultural assumption that perceivers' perspectives have epistemological authority over individuals' self-knowledge (Serano 2007). In these cases, hegemonic gender attribution practices rendered Jim, Lance and possibly the person Saulo saw "gender pretenders" (Bettcher 2007). Despite increased social awareness of trans paradigms, perceivers in cishet-dominated spaces continued to attend to outward bodily appearance when attributing gender, holding trans people accountable to biologized notions of the cisnormative gender paradigm.

Participants also observed perceivers prioritizing a record of the gender a person was assigned at birth over gender criteria reflecting the person's internal identity. A gender assignment record calls imagined infant genitals from the past into the present, maintaining the idea that gender is fixed and determined by birth genital appearance. This kind of misgendering typically happened where perceivers could access a gender assignment record, such as medical contexts. Sally, a white trans woman in her 50s, observed medical staff misgendering her roommate, also a trans woman, in this way in the hospital:

> Her doctor, (who was my doctor for a while; this is one of the things that prompted me to change doctors), and almost everyone in the hospital staff referred to her as "he," kept calling Sarah by Sarah's original legal name . . . granted, it's in the medical file somewhere – but that file should have been updated 10 years ago when she transitioned and had her surgery, you know, went through, did all the legal name change and everything. That all should have been changed and she should be referred to as "she"/"Sarah."

According to Sally, as Sarah was a longtime patient at this hospital, medical staff had access to her transition-related medical records, which documented the process of her internal identity being physically confirmed. Medical staff reportedly prioritized her pre-transition records, which listed her former gender assignment and former name, as more legitimate sources of information about her gender than more recent records, which listed legal and bodily changes. This misgendering points to the limitations of legal and bodily changes, which do not necessarily hold up in the course of interactions. Legal gender changes and transition-related healthcare are often inaccessible to trans people (Spade 2011), yet even when Sarah accessed legal and medical transition, others still apparently failed to honor her gender identity. This experience demonstrates how, despite legal gains, *intersubjective transition* is often inaccessible, due to the persistence of gendering norms, which reflect the cisnormative notion that gender is unchangeable. Sarah's experience of misgendering illustrates the priority of gender assignment record over other gender criteria, such as declared identity, gender presentation, legal documentation and, interestingly, genital appearance, in cisnormative gender attribution.

Transnormative Attention to Body Shape, Aesthetics, Pronouns, Race and Gender

In addition to being misgendered by unaware perceivers, participants also experienced being misinterpreted by those they expected to be familiar with trans paradigms and linguistic gendering practices. They perceived that these gender attributions reflected interpretations of their gender, pronouns, aesthetic style and body shape. White racial framing (Feagin 2010) also likely played a role. Examining these various components that contribute to in-group misgendering, I re-tool the notion of "transnormativity", which was originally used to describe trans people's accountability to medical standards (Johnson 2016) and assimilationist political orientations (Aizura 2006; Snorton and Haritaworn 2013). I use the term *transnormative misgendering* to describe how, despite their intentions, people regulate each other in social spaces by holding each other accountable to particular expectations for gender presentation. In particular, gender attributions based on perceptions of race, gender, pronouns, body shape and aesthetics act as regulatory practices. Such attributions reflect narrow understandings of transness. Transnormative misgendering thus describes how notions of transness shape ways of interpreting people (the term does not describe categories of people).

For example, Don, a mid-20s, Asian, self-described FTM (female-to-male) who wore "men's" clothing and long hair and had not made physical changes to his body, experienced being misread as a cis woman in queer and trans spaces as well as in cishet-dominated spaces. He explained:

> I get misgendered a lot in queer spaces even though they're advertised as trans. It reminds me of how excluded I am from this notion of transitioning, and who counts as trans, as gender nonconforming. A lot of times, to strangers, I'm not read that way, [as] being anything other than female, woman. . . . There is a lot of emphasis on looks and [there are a lot of] high energy, queer-fabulous folks in queer and trans spaces. And maybe I'm not [read as trans] in that space. You need to look a certain way to be recognized as queer. Not in terms of gender or sexuality, but the way you are socially queer.

Don concluded that looking and acting "high energy" and "queer-fabulous", which in his assessment he did not, made people legible as trans and queer in trans/queer community spaces. Although we would expect to find social actors adhering to queer trans (verbal) gender attribution practices in trans/queer spaces, Don's account suggests that dress and appearance are still important to interpreting trans and queer identities. Thus, Don perceived that social expectations of conventionally gendered aesthetics and behaviors rendered him illegible as trans. He experienced interlocutors, who he anticipated were familiar with verbal gender accomplishment practices, interpreting his appearance instead of asking him about his pronouns.

When I asked Caro, a mid-40s, gender nonconforming, genderqueer, masculine of center, light skinned person of Mexican descent what kinds of appearance signified transness, they discussed certain clothing: plaid, 90's grunge and youth fashion. They also reflected on changing trans/queer styles over time: ten years ago, it was the faux-hawk and wallet chain, but now it is the undercut and keys on a carabiner. Some of these appearance norms for transness are similar to those described by Atkins (1998) and Holliday (2001) for lesbians – what is legible as an androgynous "look".

In addition to Don's style of dress, his perceptibility as Asian and others' perceptions that he is female contribute to his "look". White perspectives that Asian men are effeminate (Chan 2001) and that Asian women are ultra-feminine (Aizura 2011) along with Don's long hair and non-use of hormones – essentially, his lack of legible gender ambiguity – may have contributed to his invisibility as an FTM within white-dominated trans/queer spaces. As described by Don and Caro, aesthetic appearance shaped gender attribution in trans/queer spaces, where interlocutors did not always communicate verbally to determine gender and/or pronouns. This illustrates the uneven integration of the queer trans paradigm into social practices.

Gender minorities who understood themselves in terms of the queer trans paradigm experienced misgendering when, according to their accounts, perceivers utilized a binary trans frame. Saulo, who described himself as queer, trans, transmasculine, genderqueer, early 20s, and European, who was on a low dose of testosterone, experienced misgendering from people he estimated were familiar with trans people. Saulo preferred to be "viewed as not masculine but boyish", and identified "strongly as *not* a man". He explained, "I embody different kinds of masculinities but not hegemonic masculinity", sharing:

> I don't like it when people assume things about me based on me using "he" pronouns. In general, it happens with people who know about trans issues. They assume that if I use "he" pronouns, based on what I look like, I am a trans man, I desire certain changes for my body, I am a man, sometimes I am straight, that I want to embody a certain type of masculinity. Happens a lot with transmasculine people who do identify as trans men.

Here, Saulo recalled being misunderstood by perceivers he estimated were familiar with trans issues. Although Saulo perceived that these interlocutors recognized him as trans and did not code him as female, misgendering in this case appeared to reflect adherence to transnormative understandings of trans identity. Similar to Don and Saulo, Lance shared how he had trouble in trans/queer communities with:

> people seeing me as not trans enough or genderqueer enough or man enough. In trans and queer communities, I've had people say things to me like, genderqueer doesn't exist, it is a bullshit political identity, you have to choose a side, and then choosing that side for me. When I first started openly identifying as genderqueer . . . one of the [trans] mentors of [my] support group, when I said I identify as genderqueer, was really dismissive,

and told me it wasn't a real identity. And he wasn't just like, that's not a real identity, you must be a trans guy, he was like, you must not be trans at all because you're too feminine, and so you're a girl. And I felt really unwelcome in the space . . . a couple years later I started dating a girl who didn't identify as genderqueer or trans . . . and she was read as much more masculine than I was, just because the way her body looked and the way she carried herself – she had smaller breasts, she was slender. And she was immediately welcomed into that space even though she didn't identify that way and didn't necessarily desire to be welcomed into that space. And that same trans mentor immediately gave her a male version of her name and started using male pronouns with her without asking.

According to Lance, the trans "mentor" denied his trans identity and dismissed his genderqueer identity, disregarding and contesting his verbal expressions of identity. Lance perceived that this mentor saw him as "too feminine" because of his body shape. Like Don, Lance experienced others interpreting him as not trans; like Saulo, Lance experienced others invalidating his genderqueer identity. The mentor also misgendered a woman he was dating, perceiving that she was trans, and not asking her to share how she wanted to be understood or referred to. This account suggests that transnormative appearance expectations, which reflect interrelated cisnormative ideas of body shape and gender identity, contributed to gender minorities' misgendering experiences.

Trans illegibility also came up for participants who experienced others reading them appropriately as their self-identified gender but inappropriately as cisgender. Eli, a light-skinned African American trans man in his early 40s who had had hormone treatment and several surgeries, described an experience where he was invited to a nightclub event exclusively for queer women and trans people. When he approached the entrance, he encountered resistance from the bouncer, who told him, "This is a queer trans space only", indicating that she was not reading him as queer or trans. He recalled:

I say, I know, my friend had invited me to come, Lindsay, who's running the entire thing. . . . So, sure enough they're giving me a hard time. So finally, I say, I am trans, and then they don't believe me. And then they start to get a little testy. She said, do you have any identification? . . . I didn't want to sit there and explain . . . so I walked out.

Eli also recalled that, "The guy in front of me read as genderqueer, looked like a trans guy", and was admitted to the party. It seems here that the bouncer excluded Eli from this trans people/queer women's space because she read him as a cis man. Eli also drew a connection between trans exclusion and racial exclusion, pointing out, "This is how gay clubs used to keep Blacks out of the club in the south . . . they'd ask how many pieces of fucking ID can you give before you didn't have any more to give, and then you couldn't get into the club." As certain "queer" adornment styles seem to signify transness within a transnormative frame, and as Black masculinities are framed as already gender deviant (Hill Collins 2004), Eli's legibility as a Black man may have contributed to his illegibility as trans in this case. Thus, in this instance, transnormative framing may have involved white racial framing and transnormative appearance norms.

Research has illustrated how trans individuals are not always welcome in LGB spaces (Doan 2007); this study shows how many are also not always recognized or welcome in *trans* spaces. Participants understood that transnormative expectations of body shape, pronouns and aesthetic appearance contributed to misgendering from interlocutors in white-dominated trans/queer spaces. Even within such spaces, where one would expect the queer trans paradigm to shape behavior, participants observed others inconsistently using linguistic gendering practices. Utilizing a queer trans paradigm would have enabled perceivers to honor participants' gender identities. White perspectives' tendency to figure bodies racialized as non-white as already gender nonconforming – as either more or less feminine or masculine than the imagined white reference point – likely contributed to participants' trans invisibility in white-dominated trans/ queer spaces. These experiences demonstrate that aesthetics that are read as signs of queer sexuality are also read as signs of queer gender – just as sexuality is read through a gendered lens (e.g., Pfeffer 2014; Namaste 1996), gender is read through a sexualized lens. The accounts of participants in the communities I investigated suggest that gender is also often read through a transnormative lens, reflecting embedded assumptions with regard to body shape, gender, aesthetics, pronouns and race. Thus, complex performances of transnormative gender attribution and accountability resulted in misgendering, exemplifying additional ways in which, as Hardie has previously shown (2006), oppressed minorities reproduce within their own ranks the dominant discourses of their oppressors.

Conclusion

This chapter responds to Hopkins' (2008) call for the development of critical geographies of bodies. Examining geographies of trans embodiment, I focus on the multiple and overlapping ways in which body appearance influences negotiations of everyday spaces. According to participants' narratives, cisnormative expectations of outward bodily appearance and record of gender assignment shapes misgendering in cishet-dominated spaces and transnormative expectations of aesthetic appearance, body shape, pronouns and race shaped misgendering in trans/queer spaces. Transnormative misgendering experiences do not reflect pervasive notions that gender is fixed, biological or determined by genital appearance at birth. However, despite transgender paradigms, which hold that gender is self-determined and should be verbally indicated and not assumed, transnormative misgendering experiences maintain notions that gender is visible on the body. Some instances of the transnormative misgendering in this study can generally be explained by the uneven distribution of specialized subcultural knowledge about queer trans conceptualizations of gender and verbal gender accomplishment practices. These practices appear to have developed within certain segments of trans and queer communities in particular urban places where such communities exist, such as the Bay Area, where I conducted this research. Therefore, although we might expect people living in the Bay Area to have been exposed to such practices, some trans and queer community members still seemed unaware of them. Regardless of the cause, both cisnormative and transnormative misgendering illustrate the persistence of biologized gender ideologies, and transnormative misgendering suggests an entrenchment of transnormative gender ideology based on the medical model of trans identity.

Examining narratives of misgendering provides a deeper understanding of how various components of gender frames take priority and/or operate together in interactions and how misgendering contributes to the complex formations of inequality that take shape in the lives of gender minorities. In addition to the repatriation of trans men into normative masculinity (Schilt 2010), misreading gender minorities as either men wearing women's clothes, women wearing men's clothes, trans men, genderqueer, cis women or cis men held gender minorities accountable to cisnormative and transnormative gender paradigms. As such, participants faced multiple kinds of gender disciplining in both cishet and trans/queer spaces. Regardless of awareness or intent, misgendering served as a mechanism of social control, invalidating internal identities and suggesting that individuals should act like, look like or be how others perceived them to be. Misgendering thus (mis)shaped the meaning produced about individuals, contributing to their marginalization.

The points this study raises and its limitations point to the need for further research. A larger sample would expand the scope and the reasonable points of comparison. Further research is needed to more directly investigate interlocutors' perceptions of gender minorities, and especially to examine how race attribution matters for gender attribution. There is a need for more research on misgendering within various trans and queer spaces and communities, and on misgendering within families and between intimates and familiars. More in-depth examination of misgendering in specific geopolitical and institutional sites is needed. Studies on the misgendering of people who do not identify as trans or gender minority, research on misperception of different identity categories – such as race, sexuality, age and ability – and different relational identities and research utilizing a deeper intersectional framework are needed. Finally, to remedy the cisnormative character of research on topics unrelated to gender diversity, we need to insist on study designs and analyses that resist and interrogate dominant gender paradigmatic assumptions, even when gender minorities are not the subject of analysis.

References

Adler, P. and P. Adler. 1987. *Membership roles in field research*. Newbury Park, CA: Sage.
Aizura, Aren Z. 2006. Of borders and homes: The imaginary community of (trans) sexual citizenship. *Inter-Asia Cultural Studies* 7(2): 289–309.

Aizura, Aren Z. 2011. The romance of the amazing scalpel: Race, affect, and labor in Thai gender reassignment clinics. In *Queer Bangkok*, edited by Peter A. Jackson. Hong Kong: Hong Kong University Press.

Angrosino, M. V. 2005. Recontextualizing observation: Ethnography, pedagogy, and the prospects for a progressive political agenda. In *The Sage handbook of qualitative research* (3rd ed.), edited by N. K. Denzin and Y. S. Lincoln. Thousand Oaks, CA: Sage.

Atkins, D. 1998. *Looking queer: Body image and identity in lesbian, bisexual, gay, and transgendered communities.* New York: Harrington Park Press.

Bettcher, Talia. 2007. Evil deceivers and make-believers: On transphobic violence and the politics of illusion. *Hypatia* 22(3): 43–65.

Bornstein, Kate. 1994. *Gender Outlaw: On men, women, and the rest of us.* New York: Random House, Inc.

Carapezza, Kirk. October 20, 2014. Mount Holyoke's new transgender policy redefines women's education. *The Huffington Post.* www.huffingtonpost.com/2014/10/20/mount-holyoke-transgender-policy_n_6015454.html.

Chan, J. 2001. *Chinese American Masculinities: From Fu Manchu to Bruce Lee.* New York: Routledge.

Doan, Petra L. 2007. Queers in the American city: Transgendered perceptions of urban space. *Gender, Place and Culture* 14(1): 57–74.

Dozier, Raine. 2005. Beards, breasts, and bodies: Doing sex in a gendered world. *Gender and Society* 19(3): 297–316.

Duggan, Lisa. 2003. *The twilight of equality: Neoliberalism, cultural politics, and the attack on democracy.* Boston, MA: Beacon.

Eversley, Melanie and Abbott Koloff. January 30, 2017. Boy Scouts of America to welcome transgender youngsters. *USA Today.* www.usatoday.com/story/news/2017/01/30/boy-scouts-america-welcome-transgender-youngsters/97268506/.

Feagin, Joe R. 2010. *The White racial frame: Centuries of racial framing and counter-framing.* Routledge: New York.

Garfinkel, Harold. 1967. *Studies in ethnomethodology.* Englewood Cliffs, NJ: Prentice-Hall.

Goffman, Erving. 1977. The arrangement between the sexes. *Theory and Society* 4: 301–331.

Hardie, A. 2006. It's a long way to the top: Hierarchies of legitimacy in trans communities. In *Trans/forming feminisms: Trans/feminist voices speak out.* edited by K. Scott-Dixon. Toronto: Sumach Press.

Hill Collins, Patricia. 1986. Learning from the outsider within: The sociological significance of black feminist thought. *Social Problems* 33: S14–S32.

Hill Collins, Patricia. 2004. *Black sexual politics: African Americans, gender, and the new racism.* New York: Routledge.

Hollander, Jocelyn A. 2001. Vulnerability and dangerousness: The construction of gender through conversation about violence. *Gender & Society* 15: 83–109.

Holliday, R. 2001. Fashioning the queer self. In *Body dressing*, edited by J. Entwistle and E. Wilson. Oxford: Berg.

Hopkins, Peter. 2008. Critical geographies of body size. *Geography Compass* 2/6: 2111–2161.

Jenness, Valerie and Sarah Fenstermaker. 2014. Agnes goes to prison: Gender authenticity, transgender inmates in prisons for men, and pursuit of "the real deal". *Gender & Society* 28(1): 5–31.

Johnson, Austin. 2016. Transnormativity: A new concept and its validation through documentary film about transgender men. *Sociological Inquiry* 86(4): 465–491.

Kelly, Reese. 2012. *Borders that matter: Trans identity management.* Dissertation, University at Albany, State University of New York.

Kessler, Suzanne and Wendy McKenna. 1978. *Gender: An ethnomethodological approach*. New York: John Wiley & Sons.

Lucal, Betsy. 1999. What it means to be gendered me: Life on the boundaries of a dichotomous gender system. *Gender & Society* 13: 781–797.

Maykut, P. and R. Morehouse. 1994. *Beginning qualitative researchers: A philosophical and practical guide*. Washington, DC: Falmer.

Meadow, Tey. 2011. "Deep down where the music plays": How parents account for childhood gender variance. *Sexualities* 14(6): 725–747.

Medin, D. L. and A. Ortony. 1989. Psychological essentialism. In *Similarity and analogical reasoning* edited by S. Vosnaidou and A. Ortony. Cambridge, England: Cambridge University Press.

Mullings, B. 1999. Insider or outsider, both or neither: Some dilemmas of interviewing in a cross-cultural setting. *Geoforum* 30: 337–350.

Namaste, Viviane. 1996. Genderbashing: Sexuality, gender, and the regulation of public space. *Environment and Planning D: Society and Space* 14(2): 221–240.

Namaste, Viviane. 2000. *Invisible lives: The erasure of transsexual and transgendered people*. Chicago: The University of Chicago Press.

Nordmarken, Sonny. 2014. Microaggressions. *Transgender Studies Quarterly* 1(1–2): 129–134.

Nordmarken, Sonny. 2019. Queering gendering: Trans epistemologies and the disruption and production of gender accomplishment practices. *Feminist Studies* 45(1): 36–66.

Nordmarken, Sonny and Reese Kelly. 2014. Limiting transgender health: Administrative violence and microaggressions in health care systems. In *Health care disparities and the LGBT population*, edited by Vickie Harvey and Teresa Housel. Lanham, MD: Lexington Books, pp. 143–166.

Norwood, Kristen. 2013. Meaning matters: Framing trans identity in the context of family relationships. *Journal of GLBT Family Studies* 9(2): 152–178.

Pfeffer, Carla. 2014. "I don't like passing as a straight woman": Queer negotiations of identity and social group membership. *American Journal of Sociology* 120(1): 1–44.

Prosser, Jay. 1998. *Second skins: The body narratives of transsexuality*. New York, NY: Columbia University Press.

Rahilly, Elizabeth. 2015. The gender binary meets the gender-variant child: Parents' negotiations with childhood gender variance. *Gender & Society* 29(3): 338–361.

Ridgeway, Cecilia. 2009. Framed before we know it: How gender shapes social relations. *Gender & Society* 23(2): 145–160.

Rose, P. 1985. *Writing on women: Essays in a renaissance*. Middletown, CT: Wesleyan University Press.

Schilt, Kristen. 2010. *Just one of the guys? Transgender men and the persistence of gender inequality*. Chicago: University of Chicago Press.

Schilt, K. and E. Windsor. 2014. The sexual habitus of transgender men: Negotiating sexuality through gender. *Journal of Homosexuality* 61(5): 732–748.

Serano, Julia. 2007. *Whipping girl: A transsexual woman on sexism and the scapegoating of femininity*. Emeryville, CA: Seal Press.

shuster, stef m. 2017. Punctuating accountability: How discursive aggression regulates transgender people. *Gender & Society* 31(4): 481–502.

Snorton, C. R. and J. Haritaworn. 2013. Trans necropolitics: A transnational reflection on violence, death, and the trans of color afterlife. *The Transgender Studies Reader* 2, Routledge, pp. 66–76.

Spade, Dean. 2011. *Normal life: Administrative violence, critical trans politics, and the limits of law*. Brooklyn, NY: South End Press.

Stone, Sandy. 1991. The empire strikes back: A posttranssexual manifesto. In *Body guards: The cultural politics of gender ambiguity,* edited by Julia Epstein and Kristina Straub. New York: Routledge.

Talbot, K. 1998–99. Mothers now childless: Personal transformations after the death of an only child. *Omega* 38(3): 167–186.

Tompkins, Avery. 2014. "There's no chasing involved": Cis/trans relationships, "tranny chasers," and the future of a sex-positive trans politics. *Journal of Homosexuality* 61(5): 766–780.

Wentling, Tre. 2015. Trans★ disruptions: Pedagogical practices and pronoun recognition. *TSQ: Transgender Studies Quarterly* 2(3): 469–476.

West, Candace and Don Zimmerman. 1987. Doing gender. *Gender & Society* 1(2): 125–151.

Westbrook, Laurel and Kristen Schilt. 2014. Doing gender, determining gender: Transgender people, gender panics, and the maintenance of the sex/gender/sexuality system. *Gender & Society* 28(1): 32–57.

Zimman, Lal. 2017. Trans people's linguistic self-determination and the dialogic nature of identity. In *Representing trans: Linguistic, legal, and everyday perspectives,* edited by Evan Hazenberg and Miriam Meyerhoff. Wellington, NZ: Victoria University Press.

10 Transitioning Through the Toilet

Changing Transgender Discourse and the Recognition of Transgender Identities in Japan

S.P.F. Dale

In discussing transgender rights, one of the most contentious issues has been one of the most basic human needs – that of using the toilet. Around the world, there continue to cases of violence against trans people based on the toilet they use, schools and institutions trying to restrict trans individuals from using the toilet of the gender they are, and discursive violence in the form of scaremongering and gender essentialism about controlling which toilet trans individuals use (Doan 2010; Gershenson 2010; Herman 2013; Namaste 1996; Patel 2017). We all need to use the toilet, but for many individuals this need has become one mired by the politics of gender, thus making the toilet a space for judgment as well as potential danger.

This chapter focuses on transgender issues in Japan. The toilet presents a macroscopic example of how gender is policed, maintained and understood in society at large, and through using the toilet as a lens this paper introduces the current legal and social context of being transgender in Japan, and brings forth the ambiguity and complexity of living as transgender. To introduce the Japanese context, I start this chapter with three recent lawsuits brought up by transwomen in Japan against their employers and, in one case, a gym. The three cases all deal with the toilet, changing rooms and transgender individuals in Japan, and reveal how transgender individuals are recognized and understood in the social and legal landscape, as well as how the toilet and changing room serve as contentious spaces.

In November 2015, a transgender woman employed at the Japanese Ministry of Economy, Ministry, and Trade sued the state for not being allowed to use the female toilet at work. The woman, in her forties, had started passing as female at work since 1998, but was not allowed to use the female toilet because legally she was still male. A superior at work had told her that if she did not intend on getting sex reassignment surgery (SRS)[1] and changing her legal gender, she should return to being a man. She was then transferred to a different section, and forced to come out as transgender to all of her coworkers.

In June 2016, a transgender woman sued her employer, Yakult Holdings. The woman, also in her 40s and legally male, received a diagnosis for "Gender

DOI: 10.4324/9781315613703-10

Identity Disorder" (GID) in 2014. She submitted this diagnosis to her superior at work, and requested her gender be changed on official documents at work. She asked to continue working using her male name, but to be able to use a changing room other than the male one at work. Her company agreed to this, but only on the condition that she first inform, in person, all of her co-workers of her transgender status.

In June 2017, a transwoman brought a lawsuit against Konami gym. In this case, she was suing them as a customer. The woman had started her gym membership as a male in 2009, and in 2014, after undergoing SRS, wrote a request to the gym to allow her to use the female changing rooms. Although she had undergone SRS, the woman was unable to change her legal gender because she was still married and had a child who was under the age of 20. Initially, the gym agreed to allow her to use the female changing room. However, after checking in with the head office the decision was reversed, and the woman was told that unless she changed her legal gender to female, she should abide by her legal male gender and dress as a man and use the male changing rooms in order to not cause inconvenience to other customers.

The above lawsuits hint at how transgender is recognized in Japan, and the first section of this chapter discusses in further detail the current legal and social framework in which transgender individuals are understood in Japan. Following this I discuss research about toilets and transgender, and contrast this with the situation in Japan, interweaving the theoretical discussion with ethnographic research that looks at the experiences of non-binary gender individuals in Japan. These empirical experiences demonstrate how the public toilet can serve as a space through which individuals negotiate their gender, but also which compels them to abide by gender norms. I argue that the public toilet is a nexus of transgender politics, and is a culmination of ideology about gender and sexuality as well as individual performance and identity. The toilet is understood as a space of transition – of discourses about transgender, as well as of transgender identity.

The Context: Transgender in Japan

In Japanese, there are presently two main terms used to refer to trans individuals – "*sei dōitsu sei shōgai*" and "*toransujendā*". The former is the Japanese translation of Gender Identity Disorder (GID) and the latter is a transliteration of "transgender". Both of these terms are relatively recent and emerged in the late 1980s (toransujendā) and 1990s (sei dōitsu sei shōgai), and prior to this there were other terms used to refer to trans individuals, some of which continue to be used today (McLelland 2004; Mitsuhashi 2003; Yonezawa 2003). The use of these terms indicates different frameworks through which to understand transgender identity, be it through a medical model or an agential-based, social constructivist one. The understanding of transgender as a medical condition has not only made it easier for trans individuals to gain social recognition, but it also created a discourse of trans individuals as pitiable and weak (Mitsuhashi 2003).

Sei dōitsu sei shōgai (henceforth referred to as GID – as it is also abbreviated in Japanese) received tremendous media attention when it was first introduced, and television dramas featuring transgender characters also started popping up shortly afterwards. Because of this development GID is more well-known than the term "transgender", although there are activists who are currently attempting to change the use of transgender terminology. There is also more awareness of transgender issues than LGB (lesbian, gay, bisexual) issues, and more policies pertaining to trans individuals.

"Sei dōitsu sei shōgai" is the term most commonly used to refer to transgender individuals, and is used in political and legal discussions about transgender and employed in official documents and policies. This term was established in 1996, when the first SRS procedure was legally conducted in Japan. The term used to translate "disorder" is "*shōgai*", which in Japanese refers to disabilities (e.g., *shintaiteki shōgai* refers to physical disabilities, and *shōgaisha* generally refers to people with disabilities). The distinction between disorder and disability in Japanese is a topic that warrants an examination on its own, but for this chapter it will suffice to say that this ambiguity has had ramification for how transgender individuals have come to be understood in Japan. It has also influenced the discussion about toilets for trans individuals, as will be discussed in the next section.

Because transgender is understood as a "shōgai" – a disability/disorder – trans individuals receive more explicit social support than cisgender queer individuals do. For example, the Ministry of Education, Culture, Sports, Science and Technology (MEXT) issued a set of guidelines for treating transgender students in schools in 2015. Later in the year it also issued a Q&A pamphlet about LGBT students (the term used being *seiteki shōsūsha* – sexual minorities). However, the majority of this pamphlet pertained to transgender issues, and same-sex attraction was hardly delved into. Another example is counseling and social support services. The national university where I used to work has a counseling center that offers support for specifically 1) students who seek general counseling and 2) disabled students. Transgender students are counted as "disabled students", and upon counseling can also discuss issues such as their gender on university records, name to be used, and other problems they may encounter on campus. However, no such support exists for cisgender queer students. Such students may use the general counseling services and receive support, but the counseling center does not explicitly advertise their services for LGB students. The organization of this counseling center is not unique, and there are many institutions similar to it across universities and institutions in Japan.

Although being transgender may be recognized as a form of disability, trans individuals do not get the same benefits and welfare services that other disabled individuals in Japan do. Individuals who are recognized as having a disability are issued with a *shōgai techō* – a "disability notebook" – which allows disabled individuals to receive specific support services. However, transgender individuals are not eligible for this.

As mentioned before, this formulation of trans identities as disorder/disability became established in the late 1990s, and prior to this SRS procedures were illegal in Japan, although they did take place underground (McLelland 2005). In Japan, any modification of one's genitals or reproductive organs is against the law because of the Maternal Protection Act, a law which was based on the Eugenics Protection Law that was created in the lead up to World War II, and which continues to exist in Japan today. The establishment of this law was meant to promote reproduction through making abortion and sterilization illegal, as well as preventing "undesirable" reproduction such as by individuals with mental disabilities (Frühstück 2003). SRS procedure became illegal because of this law, and in order to allow such procedures to be conducted in Japan legal and medical exceptions had to be established. The condition of GID was created (or one may say imported through the Diagnostic and Statistical Manual of Mental Disorders (DSM), which is also used in Japan) in order to allow transgender individuals to undergo SRS in Japan.

The medicalization of trans identities in Japan is recent compared to other countries, and Japan has a rich and vivid transgender history prior to the establishment of GID in Japan (McLelland 2005; Mitsuhashi 2003). Mitsuhashi describes the 1980s as the "golden age" of transgender in Japan, owing to how trans individuals frequently appeared in the media on variety shows, albeit as an object of entertainment. As a result transgender individuals were visible in Japan, but because there was no explicit distinction between homosexuality, cross-dressing and being transgender, there was also a lot of confusion and ambiguity about the overlapping of these identities. In terms of terminology, the neat distinction between gay and trans did not exist during this period either, although there were some individuals who identified specifically as gay rather than transgender, and vice versa. Many trans individuals were also referred to as *josōka* – as female cross-dressers. Cross-dressing communities consisted not only of transwomen, however, but also male-assigned individuals who just enjoyed cross-dressing.

The establishment of GID in Japan created a divide in these communities. Prior to this trans individuals had been viewed primarily as entertainers, but there was now suddenly a new gravitas to their identities – they had a shōgai, a condition, a disorder/disability. Mitsuhashi has described the change in public perception that she as a transwoman experienced, and says that she went from being an object of entertainment to an object of pity. For her, the former was preferable, for at least her agency was recognized. Divisions in trans communities also arose because of individuals who were for the establishment of GID, and those who were against it. Many trans individuals also started to feel the need to get diagnosed to confirm if they really were transgender. Identity was taken from the hand of the individual, and placed in the medical institution. Guidelines outlined conditions that trans individuals had to in order to be diagnosed, and during this initial period (late 1990s, 2000s) doctors also had a set idea of what it meant to be a "woman" or a "man", and the narratives provided by trans individuals needed to reflect those narrow conceptions of gender.

In 2003, it became possible for trans individuals to legally change their gender in Japan. This was due to intense lobbying by transgender activists and sympathetic cabinet members (Oe et al. 2011). The conditions for legal gender change are as follows:

1) not be less than 20 years of age
2) not be currently married
3) currently have no child who is a minor
4) have no reproductive glands or whose reproductive glands have permanently lost function; and
5) have a body which appears to have parts that resemble the genital organs of those of the opposite gender

The third requirement originally stated that one should not have any children, but was amended to "no child who is a minor" in 2008 after it was argued that "no children" was too restrictive, given how many transgender individuals had conformed to social expectations and lived heterosexual, married lives and had children. Otherwise these conditions remain as amended today, although there are many trans activists who continue to voice their opposition to them. Unfortunately, there does not seem to be much momentum at present for change. Owing to the strictness of these conditions, there are many trans individuals who choose not to change their legal gender, and who continue living and presenting as a gender different from their legal one. It is also because of this requirement that the transwoman who sued Konami gym (discussed at the start of the chapter) was unable to change her legal gender despite meeting the other requirements.

The result of various terminology and conceptual frameworks through which to understand transgender experience has also led to a lack of unity in trans communities, which can be best demonstrated through the terminology individuals use to discuss their identity. A trans activist who used to head the organization GID.jp has publicly stated that she does not identify as transgender, but as GID. There are also many who say the opposite – that they are trans, but not GID. As such, although ostensibly representing the same thing, GID and transgender can be seen as two ontological paradigms through which to understand trans identity, and although there may be overlaps they are not seen as the same by certain individuals who strongly identify with either one of the terms.

It is within this dichotomous framework and understanding of trans identities that non-binary gender identities are constructed, and have come to the fore. X-gender (*x-jendā*) is a term used to refer to non-binary gender identities in Japan, and although the term itself has been around since the late 1990s, it is only in the past five years or so that it has started gaining prominence. X-gender is frequently included in discussions of trans identities, and in discussions about the toilet. As an identity, x-gender is highly contingent and understood differently by individuals. There are x-gender individuals who understand themselves as having GID, and also those who view their identity as a political opposition to the gender binary.

As a researcher doing work on x-gender, I have received more media interest in my work more than ever before, and this can also be attributed to what is referred to as the "LGBT boom" in Japan at present – an increased interest in LGBT issues, with more coverage of these issues in major news outlets as well as in policy discussions. Possible reasons for the skyrocketing interest are the upcoming 2021 Tokyo Olympics (ensuring Tokyo is "diversity friendly" is seen as key), same-sex partnership certificate systems established in select municipalities across Japan, as well as more international attention being paid to these issues. Much of the media discussion has focused on LGBT-friendly workplaces, as more companies strive toward a work environment which welcomes individuals of different genders and sexualities, with toilets a focal point of these discussions as well.

The following sections delve into how trans individuals – specifically non-binary individuals – negotiate their gender and use of public toilets, and bring together the issues discussed before and explore how they apply to the everyday lived experiences of trans individuals as well as the relevance of the toilet as a gendered public space that regulates gender in Japan.

Transgender and the Toilet – Discussions and Experiences

"Is that for transgender people?" a young American girl asks her mother, pointing at the urinal in the female toilet. This was a conversation I overheard at Narita Airport in June 2017. The conversation highlights two issues that I will discuss in this section – the prominence of transgender and toilets as a topic of discussion (even young cisgender children are aware of it), and the cultural differences in the construction of the toilet as a public space. In discussing these issues, I will also make use of ethnographic research that I conducted with x-gender individuals in Japan as part of a larger research project that explored x-gender identity (Dale 2014). As part of this project I interviewed individuals who identified with the term "x-gender", and explored how the term has been constructed as well as how individuals use it to understand their gender and identity. Although including other trans individuals (in particular trans women, who are often the focus of debates concerning toilet use) would have been fruitful, I unfortunately do not have these data.

I focus on the cases of six individuals – Kuro, Mura, Ono, Yama, Aoi and Taka (see Table 10.1). These individuals come from different regions in Japan, but as there were no notable differences based upon geographic region, this has not been remarked upon. Identity refers to the term that individuals identify with most. Although Kuro identifies as FtM (female to male), they also identify as being non-binary. FtX refers to "female to X", and MtX "male to X", both abbreviations for x-gender. Individuals were asked what pronouns they prefer in English, and I use "they" for individuals who prefer gender-neutral pronouns. Although I will not discuss it in this chapter, pronoun use in Japanese is different than from in English, and it is possible to speak about an individual without using gendered pronouns. Language and gender are constructed

Table 10.1 Research participants

Name	Age	Identity	Legal gender	Pronoun
Kuro	Mid-40s	FtM	Female	They
Takai	Late-20s	FtX	Female	No preference
Ono	Late-20s	FtX	Female	They
Yama	Mid-20s	*Musei* (genderless), MtX	Male	They
Mura	Early-20s	MtX	Male	They
Aoi	Late-20s	FtX	Female	They

differently. Takai had no preferred pronoun, and as such I opted for the use of "they", perhaps inadvertently gendering them in doing so.

Public toilets have come to be central in discussions about gender and space, and in some countries (most notably the United States) have also become the crux of discussion about transgender rights. Most public toilets are segregated by gender, and this segregation invokes an understanding of what gender is, and how gender should be determined, as well as the purpose of gender segregation.

Past research demonstrates how the public toilet is a problematic space for transgender or gender non-conforming individuals to maneuver, and points out the social norms and expectations that persist in governing interactions in the public toilet (Barcan 2005; Browne 2007; Cavanagh 2010; Doan 2010; Gershenson & Penner 2009; Herman 2013; Halberstam 1998; Molotch 2010; Namaste 1996; Nirta 2014; Skeggs 2001). For many trans and queer individuals, public toilets are a space that provoke not only embarrassment but also fear of and actual cases of violence. Doan's phrasing of "gender tyranny" captures the sense of fear and reigning in of gender expression that transgender and gender non-conforming individuals are subject to in specific spaces (Doan 2010), and Browne's "genderism" the more subtle ways that non-gender conforming individuals are subjected to discrimination in everyday life (Browne 2007).

Genderism refers to "the (re)making of bodies and spaces through the policing of gender transgressions" (Browne 2007: 335), and with respect to the toilet, genderism is a practice that encourages/enforces individuals in act in accordance with gender norms. Acting in accordance with gender norms means different things for different individuals – it could mean trying to pass as a woman in order to use the female toilet, although one may identify as male, such as may be the case for a transman who is still legally female and/or does not have an overtly masculine physical appearance. It could also mean emphasizing one's feminine physical features when using the female toilet, as may be the case for a transwoman or a butch individual. Identity and legal gender are not the only factors that determine which toilet an individual uses, but also the interactions that one expects may occur, physical appearance and social context. It is through genderism that public toilets are maintained as highly gender normative spaces, regardless of the identities and genders of the individuals who actually make use of them.

Surveillance and Safety

Much of the discussion about transgender toilet use in western contexts has been framed as one of safety – of women and children (and the perceived threat of trans individuals using female toilets to "prey" on them), and of transgender individuals themselves, who are at risk of violence (Sanders and Stryker 2016). Much of this discussion also relies implicitly on the construction of public toilets, and how as a space they encourage surveillance. Public toilets in the contexts discussed often have a large gap between the door and the floor, allowing individuals to get a look at the feet of the person in the stall, or how they may be using the toilet (sitting/standing). Sounds produced in individual stalls are also audible, evoking interest in the producer of sounds such as a "masculine" cough.

In Japan, public toilets allow for more privacy and comfort than they may in other contexts. The gap between the door and floor is minimized, and many public toilets are constructed to allow for maximum privacy, including the masking of sounds produced in the toilet through devices such as the "*Otohime*" sound function (Matsui 2010). Surveys demonstrate that many men prefer stalls to urinals, and there is also a demand for more stalls in male restrooms (Oricon 2011). The construction of toilets is also infused with cultural ideas about the toilet as a space as well as gender roles. The urinal that the young girl pointed to in the toilet at Narita airport is not intended for transgender individuals, but rather for young boys who enter the toilet with their mothers. This construction emphasizes the role of the woman as caretaker.

Although there is more privacy in Japanese toilets, this doesn't mean that there is no surveillance, but rather that surveillance and the motivation behind it occurs differently. Some of my informants stated that they get stared at in the toilet or directly asked their gender.

Mura is male-assigned, and primarily uses the male toilet. Using the public toilet is an uncomfortable experience for them, and irrespective of their hair length or the clothes they wear (they describe themselves as preferring unisex clothes or clothes that do not specifically "gender" them) they attract stares in the men's toilet. They describe this experience as alienating and say that being stared at or being asked their gender in the toilet (which has happened to them) makes them wish that they were "normal" just so that they would not experience such feelings of shame. As far as possible they try not to go to the toilet when it is crowded, as they say that it leads to not only them but also the people around them feeling uneasy. Mura's discomfort is one that was shared by several of my informants who used the toilet of their assigned gender – the discomfort of being stared at and being asked their gender.

Mura controls their toilet use based not only on how they feel but also on how they think they will make others feel – they self-survey themselves for the sake of others. They do not describe the issue as one of personal safety, but of comfort – of not being made to feel embarrassed, of not making others feel uncomfortable through their presence. This discomfort felt by others was based on the possibility of Mura entering the wrong toilet – people asked them their gender in order to confirm that Mura knew which toilet they were using. The genderism enacted here is subtle and tacit, and operates on the understanding of a set idea of what a man or woman should look like.

A discussion in a female-to-male drag (*dansō*) magazine also demonstrates how gender non-conforming individuals can react to genderism. The female-assigned and variously identifying individuals interviewed laughed about the toilet, and stated that they would rather use the female than male toilet because it is cleaner, but some use the male toilet when in drag. One interviewee, Kazuya, said that once a child in the female toilet exclaimed, "the men's toilet is next door". Kazuya laughed and explained the situation to the child, whose mother profusely apologized in turn (Garcon Girls 2013: 53). Individuals may find their presence questioned in the toilet, but in most cases any conflict that may arise between them and other users is swiftly solved. Confrontation seldom occurs, although individuals are made to feel uncomfortable by being stared out or having their presence questioned, and the need to justify their presence.

Although there is a lack of research in Japanese, the toilet does figure prominently in autobiographies or personal accounts written by transgender individuals (e.g., Nōmachi 2009; Tanaka 2006). In these narratives, there does not exist the same level of fear or violence as one tends to find in western accounts. For example, Ray Tanaka, who identifies as genderqueer and is assigned female, describes their first time using the men's restroom as initially nerve-wrecking, not knowing if they would be able to pass or not (Tanaka 2006). However, upon successfully passing as male, this fear and anxiety instantly transformed into relief. This quick succession of feelings is one that is echoed in other narratives – fear at not being able to pass, passing or at least not being noted upon, followed by instant relief. Most visits to the toilet do not culminate in violence, and actual physical violence is rarely encountered. This is not to say that toilets are entirely "safe" spaces in Japan, but rather that there does not exist the same level of risk of physical violence as there may in other cultural contexts. It may be more accurate to say that although the toilet may not specifically be a dangerous space, it is not a welcoming one.

Negotiating Toilet Use

Genderism highlights the expectations people hold toward gender, and who they expect to see inhabiting gendered space. These expectations also play into how trans individuals negotiate their own toilet use.

Unlike Mura who uses the toilet of their assigned gender, Kuro decides which toilet to use based on the gender they can pass as. Kuro is assigned female, and has been using testosterone for the past 20 years. They have not had any reconstructive surgery (although they do bind their breasts), pass as male in everyday interactions and also grow facial hair. At present, Kuro uses only the male restroom, and says that their masculine body frame and appearance made it difficult for them to continue using the female restroom. For them, the issue is not one of which toilet they would like to use, but which toilet they can use, and Kuro in fact says that ideally they would prefer to use female toilets because they tend to be cleaner. The choice of which toilet to use is not entirely up to the individual, but also how they perceive of others perceiving them. For Kuro, their male appearance effectively closed the door of the female toilet to them. Kuro could still venture into the female toilet if they chose to, but have decided against doing so for the sake of "convenience", to avoid being called out by someone who might question their presence there. Given that Kuro easily passes as male, this proves to be the most hassle-free and painless solution – nobody bats an eyelid, nobody questions their belonging to that (male) space.

Yama is male-assigned, and says that their appearance fits with their assigned gender. They say that there is nothing incongruous about them entering the male restroom and that they do not stand out or attract stares in any way. However, since coming to recognize themselves as x-gender they say that ideally, they do not want to use the male restroom, and as far as possible try to use the gender-neutral or disabled restroom (which are often gender-neutral) where there is one. They do admit that in many cases they use the male restroom because they feel it is inappropriate to use the disabled restroom given that they are able-bodied.

Unlike Mura who gets stared at in the toilet for not sufficiently passing as male, Yama does not attract any attention at all, and their sense of discomfort stems rather from feeling like they do not belong there, rather than other people conveying that to them (through staring, directly asking). The choice not to use the male toilet is not caused by the discomfort caused from others, but rather that of personal discomfort and, like Kuro, self-policing. The self-policing is not of performance, but rather of maintaining the gender binary – Yama does not identify as male and as such feels that they *should not* use the male toilet. Their gender identity has orientated their toilet choice. For Kuro, on the other hand, it was their gendered appearance and performance that orientated them. These orientations also demonstrate how these individuals understand gender segregation – for Yama it was based on gender identity, Kuro on performance and Mura on assigned gender.

Ono, like Kuro, is female-assigned and also binds their breasts. However, whereas Kuro passes primarily as male, Ono manages a complex web of social relationships, and has social circles where they pass as male, and others as female. Both groups are unaware that Ono also passes as another gender, and the social circle where they pass as male (which consists entirely of heterosexual cisgender men) is completely unaware of their assigned gender. Ono not only decides which toilet to use depending on who they are out with but also says that managing these dual identities is a difficult task.

> When I'm passing as male I use the male toilet. But, of course that's because I'm passing as male. Usually I use the female toilet. More recently I use the unisex toilet, but where I have to choose either female or male I choose the female toilet. But what is scary is that I don't know when I might meet someone who knows me as male, so I have to be really careful about which toilet to choose.

Ono's appearance is ambiguous enough to pass as both female and male, and this provides them greater leeway in orientating their way around restrooms. They say that they only use the male toilet when they are with members of their male group. Being with a group of men who recognizes them as "one of them" also legitimizes their presence in the men's toilet, and other users of the toilet do not pay attention to them. In Ono's case, the issue of which toilet to use is not so much an issue of how the other users of the toilet may respond to them (as it was primarily for Kuro), but rather how the people they are with recognize them. Ono uses the toilet that they are expected to use by their peers. They admit that this is in fact very stressful, as being seen entering a specifically gendered toilet could disrupt their carefully managed social relationships and performances. Being spotted entering or exiting the female toilet by a member of the group that knows them as male is a frightening prospect – they would lose their place in this all-male group, and no longer be recognized as "one of the guys".

So long as an individual "passes" as the gender of the toilet they are entering, there is no friction, no disturbance. Takai is in their mid-twenties and is female-assigned. Takai says that they have no desire to pass as male, but on occasion gets read as male nevertheless. Although they do not get stared at too often nowadays, it used to be more of an issue when they were younger.

> When I get mistaken [as a male] in the toilet I just ignore it. But, recently that doesn't really happen. Long ago, perhaps, when my skin was really tan [. . .] (that) played a role. And my hair was really short, like a guy, it was so short it would stand. And I also played sports. It wasn't intentional, but I got mistaken [as a male].

Takai notes that their appearance as a teenager matched the expectations of what a teenage boy would look like – heavily tanned (from playing sports outside), and a closely cropped haircut. It was not their intention to pass as male, but something that just happened. Although this was a slight inconvenience for Takai, they say that when they do get mistaken as a male they ignore it, and are aware of their own right to use the female toilet, and assertively hold ground.

Takai also makes use of the male toilet on occasion, in particular when there is a long line for the female toilet. For them, the issue is one that appears as matter of fact – if there is a long queue for the female toilet but hardly anyone in the male, why wait? When they use the male toilet, they also intentionally modify their appearance – if they are wearing a hoodie they put their hood on, and hunch their shoulders to assume a masculine stature. As far as possible they say that they prefer using the female restroom because it is cleaner, and although they do get stared at on occasion this is something they laugh about and take in their stride.

Like Mura, Takai's default toilet to use is that of their assigned gender, and they only use the male toilet when it is more convenient to do so. Unlike Mura, whose experiences in the male toilet also fed into their discomfort with their assigned gender, Takai's experiences did not lead them to experience any specific discomfort. Takai also recognized that as a female-assigned individual, they had just as much right to use the female toilet as other users, whereas in Mura's case this right was not one they asserted, nor felt was necessarily theirs.

Aoi is female-assigned and presently identifies as FtX. However, prior to this they had also had a period of identifying as FtM, and during this period they intentionally tried to pass as male. Aoi says that when they identified as FtM they made it a point to start using the men's toilet, as they felt it incumbent upon themselves to prove their masculinity.

> [When I identified as FtM] I wanted to become more manly, so I really wanted to use the male toilet. I wanted my manliness, my masculinity, to be acknowledged, just for myself even.

Although they describe themselves as feeling nervous the first time they used the male restroom, they gradually got used to using it. Now, however, they say that they use primarily the female toilet, and only use the male in cases of emergency. Like Takai, Aoi attempts to perform as male when using male space – hunching their shoulders or wearing a hygiene mask, for example, to appear more male.

In Aoi's case, we can see how their identity as well as assigned gender influenced their choice of toilet. Recognizing themselves as female-assigned, they primarily make use of the female toilet, but also use the male toilet on occasion. Their identity as x-gender is not necessarily what allows them this flexibility in toilet use, but rather it is their past experience in utilizing this space that makes it accessible to them. Aoi also describes specifically using the male toilet when they identified as male as an important aspect of having their manliness recognized. They no longer felt this need, however, when their identity moved towards being x-gender.

Gender Neutral Toilets, LGBT-Specific Toilets

Individuals such as Yama and Ono state a preference for gender-neutral toilets. Although trans individuals may in some cases prefer gender-neutral toilets (Lixil 2016; Porta et al. 2017), there are also cases where they are restricted to doing so (Cavanagh 2010; Jones et al. 2016; Whittle et al. 2007). As Huesmann notes, there are three different kinds of gender neutral toilet designs – "1) that found in restrooms in airplanes, trains and buses, 2) the disabled restroom and 3) diaper changing stations" (Huesmann 2016: 544–545). In Japan, all three cases exist, but in some spaces, disabled restrooms and diaper changing stations have become more explicitly gender neutral, and in some cases, LGBT specific.

In Japanese media, the term used to refer to gender-neutral toilets is "LGBT toilets". This further ambiguates the distinction between trans and gay individuals, and also supports the assumption that all LGBT individuals are not gender conforming, as well as that LGBT individuals desire to use gender-neutral toilets. The plan to install more gender-neutral toilets ahead of the 2021 Tokyo Olympics was reported as "toilets that are kind to LGBT individuals" (Massaki 2017: np) for example, and a survey conducted by LIXIL about trans experiences in the toilet was reported by another newspaper as "LGBT – over half have 'toilet stress'" (Mainichi 2016: np). Despite both cases referring primarily to trans individuals, they are reported as LGBT issues, confounding and oversimplifying the complexity of identities in the acronym. "LGBT" is used for brevity, as it takes up less characters than "transgender" would. It is also a buzzword at the moment given the media attention to LGBT issues. As these headlines also demonstrate, there are currently movements to install more gender-neutral toilets across Japan, and one of the main motivating factors for this is the Olympics, hinting at economic and international influence.

Universal, gender-neutral toilets have existed in Japan for the past few years. These toilets are designed for all individuals, and are single-stall toilets which can fit a wheelchair, or two people. They also usually have diaper-changing facilities. They are as such not only gender-neutral but cater to other needs as well.

In 2015, Shibuya ward in Tokyo became the first area in Japan to provide a form of legal recognition of same-sex (in terms of legal gender) partnerships. This partnership system enabled same-sex couples to apply for a partnership

Figure 10.1 Toilet sign with a figure in a rainbow "half-skirt"
Source: Author's image

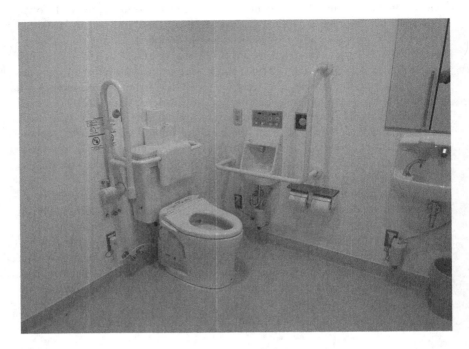

Figure 10.2 Interior of single-stall gender neutral restroom

Source: Author's image

certificate, which would grant them the same rights that married heterosexual couples have. Since then Shibuya ward has started taking into consideration other LGBT issues, including the toilet. In November 2015 the Shibuya ward office revealed new symbols for the universal toilets in their office buildings. These designs also became the subject of a Twitter debate that took place shortly after (Togetter 2016). Much of this discussion had to do with a half-skirted figure wearing a rainbow outfit, situated between figures symbolizing "man" and "woman".

Individuals who were supportive of the new symbol described it as "fantastic", and as an important act in making LGBT people more visible. However, those against it mostly found the symbol problematic, and to offer a confused representation of a transgender individual. The toilet in question is also a single-stall restroom situated next to the female and male toilets, and as such it was feared that by entering this restroom, individuals may inadvertently be "outing" themselves.

Despite the opposition from members of the trans community, public toilets utilizing this symbol have increased in recent years. Hotel Grandia in Kyoto, which promotes itself on various tourism sites as "LGBT-friendly", has also been making use of it, as have other private establishments which tout themselves as LGBT-friendly. Don Quijote, a popular discount chain store in Japan,

Figure 10.3 Gender-neutral bathroom sign with a rainbow flag
Source: Author's image.

made the news in May 2017 when they announced gender-neutral restrooms in their flagship store in Shibuya ward (*Japan Times* 2017). Articles about the toilets in Japanese discussed the signaled "all gender" toilet as "LGBT toilets". The toilets also used the half-skirted figure.

Gender-neutral toilets in Japan are always single-stall toilets and often are not just gender-neutral but also barrier-free, meaning individuals with physical

Figure 10.4 Men, all gender, women sign
Source: Author's image.

disabilities or individuals who need diaper changing facilities also utilize them. As Mura said, although they want to use gender-neutral toilets, the fact that these toilets are also intended for individuals with physical disabilities makes them question their own priority in doing so.

Conclusion

The three cases at the start of this paper and the local media attention they garnered demonstrate how transgender issues are gaining more prominence in Japan, and how the toilet is a focal point of this debate. However, the discussion taking place in Japan is remarkably different from that in western countries, and the focus is not on safety, but rather on comfort. Despite this difference, genderism and the expectation to abide by gender norms govern this space, and have the potential to lead to the alienation of trans individuals. The alienation of trans individuals occurs not only because of feelings incurred in the toilet but also because of signage and the discursive medium through which this discussion is taking place. There is a risk that trans individuals may be relegated to using universal toilets, which also prevents them from being recognized as women or men. At the same time, the ambiguity in terminology used to refer to gender-neutral toilets also supports false assumptions about LGBT individuals, and maintains an understanding of LGBT as "other".

The construction of the toilet as a space and the relative lack of violence in Japan have allowed some individuals more freedom in negotiating which toilet to use. Although some individuals experience discomfort at their gender being questioned, some also brush this off. The right that an individual feels they have to use a specific toilet is highly contingent.

The discussion about transgender and toilets needs to move beyond the current restrictions of maintaining male/female-gendered spaces, and seek more inclusive solutions, such as multi-gender toilets which all individuals can easily make use of (Sanders and Stryker 2016). In Japan, the issue should not only be about comfort but also about a less exclusionary form of inclusion that moves beyond overemphasizing difference.

Note

1 I use the term sex reassignment surgery (SRS) as it corresponds to the Japanese terminology, *seibetsu tekigō shujutsu*.

Bibliography

Barcan, R. (2005) "Dirty Spaces: Communication and Contamination in Men's Public Toilets." *Journal of International Women's Studies*, 6(2), pp. 7–23.

Browne, K. (2007) "Genderism and the Bathroom Problem: (Re)materialising Sexed Sites, (Re)creating Sexed Bodies," *Gender, Place & Culture: A Journal of Feminist Geography*, 11(3), pp. 331–346.

Cavanagh, S.L. (2010) *Queering Bathrooms: Gender, Sexuality, and the Hygienic Imagination.* Toronto: University of Toronto Press.

Dale, S.P.F. (2014) "Mapping X: The Micropolitics of Gender and Identity in a Japanese Context." PhD diss., Sophia University.

Doan, P. (2010) "The Tyranny of Gendered Spaces – Reflections from beyond the Gender Dichotomy." *Gender, Place & Culture: A Journal of Feminist Geography*, 17(5), pp. 635–654.

Garcon Girls (2013). Tokyo: Umade.

Gershenson, O. (2010) "The Restroom Revolution: Unisex Toilets and Campus Politics." In *Toilet: Public Restrooms and the Politics of Sharing.* Edited by H. Molotch and L. Norén. New York: New York University Press, pp. 191–207.

Gershenson, O. and Penner, P (2009). *Ladies and Gents: Public Toilets and Gender.* Philadelphia: Temple University Press.

Frühstück, S. (2003) *Colonizing Sex: Sexology and Social Control in Modern Japan.* Berkeley: University of California Press.

Halberstam, J. (1998) *Female Masculinity.* Durham: Duke University Press.

Herman, J.L. (2013) "Gendered Restrooms and Minority Stress: The Public Regulation of Gender and its Impact on Transgender People's Lives." *Journal of Public Management & Social Policy*, pp. 65–80.

Huesmann, M. (2016) "Transgressing Gender Binarism in the Workplace? Including Transgender and Intersexuality Perspectives in Organizational Restroom Policies." In *Sexual Orientation and Transgender Issues in Organizations: Global Perspectives on LGBT Workforce Diversity.* Edited by Thomas Köllen. Vienna: Springer, pp. 539–552.

Huffington Post (2015) "Sei dōitsu sei shōgai no keisanshō shokuin, kuni wo teiso." November 13. Available at: www.huffingtonpost.jp/2015/11/13/gid_n_8553222.html (Accessed: 15 August 2017)

Itani, S. (2011) "Sick but Legitimate? Gender Identity Disorder and a New Gender Identity Category in Japan." *Sociology of Diagnosis: Advances in Medical Sociology*, 12, pp. 281–306.

Japan Times (2017) "Don Quijote Says New Individual Restroom Stalls are for Everyone Regardless of Gender, Orientation." May 15. Available at: www.japantimes.co.jp/news/2017/05/15/national/social-issues/don-quijote-says-new-individual-restroom-stalls-everyone-regardless-gender-orientation/ (Accessed: 16 August 2017)

Jones, T., Smith, E., Ward, R., Dixon, J., Hillier, L. and Mitchell, A. (2016) "School experiences of transgender and gender diverse students in Australia." *Sex Education*, 16(2), pp. 156–171.

Kirk, M. (2016) "Japan is Using Luxurious Public Toilets to Encourage Women to Join the Workforce." *Business Insider*, October 23. Available at: www.businessinsider.com/japan-is-using-public-toilets-to-encourage-women-to-join-the-workforce-2016-10 (Accessed: 17 August 2017)

Lixil (2016) "Seiteki mainoritei no toire mondai ni kansuru web chōsa kekka." Available at: http://newsrelease.lixil.co.jp/user_images/2016/pdf/nr0408_01_01.pdf (Accessed: 15 August 2017)

Mainichi Shinbun (2016) "LGBT hansū 'toire sutoresu'." April 17. Available at: https://mainichi.jp/articles/20160417/ddm/013/040/006000c (Accessed: 20 August 2017)

Massaki, K. "LGBTra ni yasashii toire – Tokyo gorin ni muke to ga keikaku." *Asahi Shinbun*, February 26. Available at: www.asahi.com/articles/ASK2J7JYSK2JUTIL067.html (Accessed: 18 August 2017)

Matsui, Y. (2010) "Masking Toilet Noise May Date Back to Edo." *The Japan Times*, February 11. Available at: www.japantimes.co.jp/news/2010/02/11/national/masking-toilet-noise-may-date-back-to-edo/#.WaXibdFpwh4 (Accessed: 15 August 2017)

McLelland, M. (2004) "From the Stage to the Clinic: Changing Transgender Identities in Post-war Japan." *Japan Forum*, 16(1), pp. 1–20.

McLelland, M. (2005) *Queer Japan from the Pacific War to the Internet Age*. Oxford: Rowman & Littlefield Publishing.

Ministry of Education, Culture, Sports, Science and Technology (2015a) *Sei dō itsu sei shōgai ni kakawaru jidō seito ni taisuru kime komaka na taiō no jisshi ni tsuite*. Available at: www.mext.go.jp/b_menu/houdou/27/04/1357468.htm (Accessed: 15 August 2017)

Ministry of Education, Culture, Sports, Science and Technology (2015b) *Sei dō itsu sei shōgai ya seiteki shikō·sei jininn ni kakawaru, jidō seito ni taisuru kime komaka na taiō nado no jisshi nitsuite (kyōshokuin muke)*. Available at: www.mext.go.jp/b_menu/houdou/28/04/__ics-Files/afieldfile/2016/04/01/1369211_01.pdf (Accessed: 15 August 2017)

Mitsuhashi, J. (2003) "Nihon toransujendā ryakushi." In *Toransujendarizumu sengen*. Edited by Izumi Yonezawa. Tokyo: Shakai hihyōsha, pp. 96–129.

Mitsuhashi, J. (2009) "Toransujendā wo meguru sogai/saika/sabetsu." In *Sekushuariti no tayōsei to haijō*. Edited by Yoshii Hiroaki. Tokyo: Akashi shoten, pp. 161–191.

Molotch, H. (2010) "Learning from the Loo." In *Toilet: Public Restrooms and the Politics of Sharing*. Edited by Harvey Molotch and Lauren Noren. New York: New York University Press, pp. 1–20.

Namaste, V. (1996) "Genderbashing: Sexuality, Gender, and the Regulation of Public Space." *Environment and Planning D: Society and Space*, 14, pp. 221–240.

Nijiiro Diversity (2017) "Seiteki mainoritei to toire fooramu – anshin, kaiteki no toire kankyō wo mezashite." July 10. Available at: http://nijiirodiversity.jp/%E6%80%A7%E7

%9A%84%E3%83%9E%E3%82%A4%E3%83%8E%E3%83%AA%E3%83%86%E3%82%
A3%E3%81%A8%E3%83%88%E3%82%A4%E3%83%AC%E3%83%95%E3%82%A9%E
3%83%BC%E3%83%A9%E3%83%A0-%EF%BD%9E%E5%AE%89%E5%BF%83%E3%
83%BB%E5%BF%AB/ (Accessed: 15 August 2017)

Nirta, C. (2014). Trans subjectivity and the spatial monolingualism of public toilets. *Law and Critique*, 25, pp. 271–288.

Nōmachi, K. (2009). *Okama dakedo OL yattemasu*. Tokyo: Bungeishunju.

Oe, C., Torai, M., Kamikawa, A with Fujimura-Fanselow, K. (2011) "Dialogue: Activists and politicians on sexuality." In *Transforming Japan*. Edited by Kumiko Fujimura-Fanselow. New York: The Feminist Press, pp. 177–196.

Oricon Kyaria (2011). "Kōkyō toire he no yōkyū, dansei ha 'kanzen koshitsuka' he no koe ga tasuu." November 10. Available at: http://career.oricon.co.jp/news/2003582/full/ (Accessed: 15 August 2017)

Patel, N. (2017) "Violent Cistems: Trans Experiences of Bathroom Space." *Agenda*, 31(1), pp. 51–63.

Porta, C., Gower, A.L., Mehus, C.J., Yu, X., Saewyc, E.M., and Eisenberg, M.E. (2017) "'Kicked Out': LGBTQ Youths' Bathroom Experiences and Preferences." *Journal of Adolescence*, 56, pp. 107–112.

Sanders, J., and Stryker, S. (2016) "Stalled: Gender-neutral Public Bathroom." *South Atlantic Quarterly*, 115(4), pp. 779–788.

Sankei News (2015) "'Joshi toire shiyō kinshi ha sabetsu' sei dōitsu sei shōgai no keisanshō shokuin ga kuni nado teiso tōkyō chisai." November 13. Available at: www.sankei.com/affairs/news/151113/afr1511130031-n1.html (Accessed: 15 August 2017)

Skeggs, B. (2001) "The Toilet Paper: Femininity, Class and Misrecognition." *Women's Studies International Forum*, 24(2–3), pp. 295–307.

Tanaka, R. (2006) *Transgender Feminism*. Tokyo: Impact Publishing.

Teague, M. (2016) "North Carolina's Transgender Bathroom Battle: What Sparked it, and What's Next." *The Guardian*, May 14. Available at: www.theguardian.com/us-news/2016/may/14/north-carolina-bathroom-bill-hb2-transgender-rights (Accessed: 15 August 2017)

Thorn, R. (2016) "Why Toilets are a Battleground for Transgender Rights." *BBC News*, June 8. Available at www.bbc.com/news/uk-england-36395646 (Accessed: 15 August 2017)

Togetter (2016) "Reinbō Ashura danshuku? Shibuya ku kuyakusho no 'dare demo toire' ni fuhyō." November 11. Available at: https://togetter.com/li/898764 (Accessed: 23 August 2017)

Watanabe, K. (2017) "'Ikiteite yoin da to omoeta' sei dōitsu sei shōgai no jimu riyōsha, konami to wakai seiritsu." *BuzzFeed News*, June 19. Available at: www.buzzfeed.com/jp/kazukiwatanabe/20170619?utm_term=.mg8wDJwQR#.wrB2ZE21l (Accessed: 16 August 2017)

Whittle, S. Turner, L., and Al-Alami, M. (2007) *Engendered Penalties: Transgender and Transsexual People's Experiences of Inequality and Discrimination*. Available at: www.ilga-europe.org/sites/default/files/trans_country_report_-_engenderedpenalties.pdf (Accessed: 15 August 2017)

Yonezawa, I. (2003) "Toransujendā gairon." In *Toransujendarizumu sengen*. Edited by Izumi Yonezawa. Tokyo: Shakai hihyōsha, pp. 13–40.

11 Contested Identities

Cisgender Women in Trans Relationships and the Politics of Naming

Avery Brooks Tompkins

Riki Wilchins (2004) argues that the value placed on language in some cultures to stand in for the "real" has very particular effects on sexual and gender identities. As sexual and gender identities often play intricate roles in our intimate relationships with others, as well as our potential memberships in various communities, the power of a binary system of language around sexual and gender identities to erase lived experience and identity is highly problematic. This chapter presents some of the concerns around sexual identity categories for cisgender women with trans[1] partners who were assigned female at birth, while also recognizing the importance of these categories for many individuals, both personally and politically. I show how words such as "lesbian", "bisexual", and "straight" – even "queer" and "pansexual" – fail to provide adequate descriptions of identity for many of the cisgender women whose stories are shared here. These terms might help to define *individual* sexual identity, but they fail to account for the ways that gender and sexuality are also *relational* identities (see Sanger 2010, 2013; Tompkins 2011, forthcoming; Whitley 2013) that indicate our connections with intimate others.

As there is no widely used language for sexuality that takes trans identities into account, trans identities are often made invisible through the currently available labels for sexuality. What terms are cis partners using to describe their sexual identities and how are these words operating in the context of their relationships? How might cis people talk about their trans partners in ways that affirm both individuals in the relationship? How does one negotiate their own sexual identity while still identifying within their relationship? I consider these questions by examining how gendered language is related to issues of (in)visibility around sexuality, and I argue that binaries in language impose limits around sexual identity that erase the trans specificities of a relationship. Further, I posit that there is deep policing around sexual and gender identity categories that affect the ways that cis women partners name their sexualities and describe their relationships to others. This chapter also illustrates how some partners (re)define identity terms while arguing that none of the available options for sexual identity are able to seriously take trans identities into account.

Salvador Vidal-Ortiz (2002) argues that we cannot separate sexuality from gender because sexual orientation requires identification of gender identity

DOI: 10.4324/9781315613703-11

in order to make sense. For example, "straight" and "gay" require gender to be defined in binary terms in order to make sense. In other words, "gender is sexual and sexuality is gendered" (Vidal-Ortiz 2002: 182). Jason Cromwell argues that trans people "queer the binaries" of identity: "[B]y 'queering the binaries' I mean that they are peculiar, seem bizarre, and spoil the effectiveness of categories" (Cromwell 2006: 510). That is, when trans people (and their partners) use binary identity categories such as "lesbian" or "straight", these categories *become queer* in that social norms around who can claim these categories, and who these categories are presumed to describe, are challenged. While challenging binary language constructs around gender and sexuality is important and necessary to a queer politics, as Gayle Rubin argues:

> Our categories are important. We cannot organize a social life, a political movement, or our individual identities and desires without them. The fact that categories invariably leak and can never contain all the relevant 'existing things' does not render them useless, only limited. . . . We use them to construct meaningful lives, and they mold us into historically specific forms of personhood.
>
> (2006: 479)

Further, although some individuals have denounced identity-based politics in favor of affinity-based groups and social justice organizations (see Green 2006; Phelan 2004; Valentine 2007; Wilchins 1997), the fact remains that identity-based communities are still safe havens for many LGBTQ people and that identity continues to be the basis for a significant portion of large-scale organizing.[2] But how can one find a community of similar people when there is very little language with which to accurately describe one's identity or experience?

A modest body of literature has examined the relationships and experiences of cis women who partner with people on the FTM spectrum. Some of this work has considered identity issues in relation to transgender, but analyses vary depending on the disciplinary background of the researcher. For example, Brown (2005), Nyamora (2004) and Mason (2006) focus on psychological stage models of grief, loss and caregiver burden for cis women when a partner transitions. However, Brown's newer work (2009, 2010) combines a clinical discussion with social analyses around issues of sexual identity renegotiation and sexual intimacy, which is often lacking from other psychological literature. Sociologists have also contributed to researching relationships involving trans individuals. For example, research has looked at trans people's experiences of relationships (Hines 2007); power relations, intimacy and governance in cis/trans relationships (Sanger 2010, 2013); identities, bodies and family life for cis women in cis/trans relationships (Pfeffer 2009, 2010, 2017); and femmes engaging in gender labor with FTM partners (Ward 2010). Most of this work, however, has focused on the *relationship itself* in a cis/trans partnership. My own work takes a slight departure from this focus, to instead examine the experiences of identity, community and trans activism for cis partners in relation to the broader social world and social structures, as opposed to focusing on experiences within the relationship itself (Tompkins 2011, 2014, forthcoming).

Theory/Methods

This chapter follows a queer sociological perspective regarding identity and language. Unfortunately, many queer theorists, in their deconstructionist endeavors, have ignored the concrete ways that "queer" is deployed as an identity that is connected to political and collective movements in favor of a "politic [that] becomes overwhelmingly cultural, textual, and subjectless" (Gamson 1996: 409). Sociology pushes queer theory in a more social direction grounded in interactions, instead of relying on textual and cultural analyses alone (Tompkins 2011). In terms of identity, as Seidman notes:

> the aim is not to abandon identity as a category of knowledge and politics but to render it permanently open and contestable as to its meaning and political role. In other words, decisions about identity categories become pragmatic, related to concerns of situational advantage, political gain, and conceptual utility.
>
> (1996: 12)

Further, as Valocchi argues, using a queer sociological perspective recognizes how "individuals claim certain identities even as they undercut these claims through their practices and their (sometimes unstable) desires and subjectivities" (2005: 767).

The methods used in this project were largely informed by postmodern theories and intersecting queer theoretical viewpoints that have moved postmodern methodologies in more politically grounded directions. As Joshua Gamson has argued, queer theory has allowed scholars to consider new areas of inquiry and new ways of inquiring. It pushed "the postmodern moment in qualitative inquiry" into the study of sexualities (Gamson 2000: 354), and Valocchi (2005) directs us to ethnography as the method of choice for projects informed by queer theory.

Between 2008 and 2010, I conducted postmodern ethnographic research (see Dicks et al. 2006; Fontana 2003; Hookway 2008; Murthy 2008; Richardson 1988), focusing on cisgender women partnered with trans people who were assigned female at birth.[3] My call for participants specified the potential interview population as "cisgender (non-trans) people who have/had partners that were assigned female at birth but who do not identify as female/woman". Though I intended this project to be fairly open in terms of identity for both partners (i.e., inclusive of cis women and men with partners assigned female at birth who identify as anything other than cis), those who responded to the call for participants were all cis women.[4] The research for this chapter blends traditional ethnographic data from interviews and observations with digital ethnographic data from blogs and YouTube videos, to analyze stories of identity from white cisgender women who are partnered with trans people on the FTM spectrum.[5] Interviews were conducted with 18 white cis women aged 18–29 (mean = 24.1) from the United States and Canada through various means (in person, via email, and/or via messaging programs) due to geographic distance. Ninety-two YouTube vlogs[6] were transcribed from two YouTube channels that focused on the experiences of cis women with partners on the FTM spectrum. Vlogs were made by cis partners for other partners, trans community members and allies.[7] As opposed to researcher-led interviews and participant observation in more traditional ethnographic practice, these YouTube videos exist as a kind of "auto-interview" (see Boufoy-Bastick 2004) where a person is both the interviewer and the interviewee.

(In)visibility: Being Read as Straight

When trans people transition, one thing that may occur is a shift in pronoun usage since pronouns tend to act as social cues for someone's gender. One of the issues that Kate[8] brought up in a blog post that she shared with me was that when she talks about her partner, her identity gets erased as soon as she uses "he" to mention him. She wrote about how a shift in language has affected her own visibility as a lesbian:

> I'm starting to feel uncomfortable that none of these people really know who I am. Not that I'm afraid of telling them I'm gay, or that I'm seeing someone trans. It's just difficult to explain, and even more difficult to work into a conversation. When I was seeing a girl, all I had to say to new acquaintances was "my girlfriend works at such and such" or "me and my girlfriend went to the cinema." Immediately they would know and it wasn't a big announcement. Now, bringing [my partner] into the conversation immediately marks me as straight, even though I call him "my partner," the dreaded pronoun comes along soon enough. I've always been one to say I don't care what people think of me. On the other hand, I feel like I'm in the closet.

When Kate's partner was using "she" as a pronoun, Kate felt that outing herself was much simpler. However, switching pronouns for her partner has meant that Kate is now viewed as straight by others. Renee also discusses pronouns and coming out in relation to her partner, Taylor:

> There is always that issue with me identifying as a lesbian and meeting someone for the first time or, you know, like, how do I disclose or get the point across that I'm a lesbian, I'm a person who's attracted to women, but I'm going to be referring to my partner as "he," just so you know, but that doesn't mean I'm straight? Like, how do you get the average, every day, not-queer-conscious person to process that situation through their head?

Like Kate, Renee found it difficult to explain her relationship. Using "he" to refer to her partner also invited the possibility of confusion from others about whether she is straight, which ultimately led to her feeling like she needed to explain everything in order to stay true to her own identification as a lesbian. Being read as straight was also an issue for Tina, who says in her vlog about queer visibility:

> It makes me really uncomfortable when I'm perceived as straight because I absolutely do not think of my relationship with [my partner] as a straight relationship, I think of it as a queer relationship. Even if I was with a bio man, I could never have, like, a straight relationship with him. That kind of normative expression of gender doesn't really fly with me I guess.

Loss of queer visibility for some cisgender partners is something that came up in conference workshops as well. I facilitated a conference workshop in 2008 that focused on how trans people can support their partners through a transition. Of the approximately 70 attendees, most were cis/trans couples. As I listened to small-group conversations, I realized that every group had at least one discussion about how to maintain a comfortable level of queer visibility for both people in the relationship. Several cis women partners were afraid they would lose visibility in the future (especially if their partner started testosterone and was consistently read as male), and other people talked about how to gain back visibility they felt was already lost. One of the complexities around visibility for many cisgender partners is balancing their own desire for visibility with their trans partner's potential desire to be stealth or to not be seen as queer.[9] However, not *all* cisgender partners desire to be read as lesbian or queer, or even care about it. In her vlog, Faith says:

> If I had queer visibility I pretty much lost it because I know that people perceive [my partner] and I as a straight couple. Which again, is perfectly fine with me, I don't care – I know he likes it, sometimes.

While sexual identity is discussed in different ways by participants, it's important to note just how much the narratives included issues of visibility. The trope of "the closet" informs this: we are expected to "come out" of hiding and make our sexualities known (Sedgwick 1990). It is generally considered to be a time of celebration when we come out, and we are congratulated for it – even if we are subject to negative consequences for doing so, such as losing friends, family and/or employment, among other things. When considering the closet, it's not surprising that issues of queer visibility are fairly important to the majority of cis women in my work, many of whom were already involved in LGBTQ communities before meeting their trans partner. As Michael Brown points out, "coming out or staying in the closet is usually materialized in the form of a speech act" (2000: 29). The performativity of language and the speech acts required to come out produce difficulties for partners who are attempting to negotiate being read as straight while identifying within, or in relation to, LGBTQ communities. Pronouns were used as a way to come out for some participants in the past, but with a partner's transition and a shift in pronoun use they simply aren't enough to make oneself visible as queer anymore. Brown says that, "by remaining silent, by not telling one's sexual story, that which is known to the self remains unknown to others: heteronormative power is exercised once again" (2000: 44).

But some cisgender partners are not *exactly* interested in remaining silent and issues of visibility are difficult when more than one person is involved in making queerness visible. As the next section of the chapter addresses, some partners are reclaiming and redefining identity terms in ways that make sense for them in order to gain or keep some degree of visibility. Other partners are simply refusing labels for sexual identity altogether, a silence that is, perhaps, as queer as actually claiming "queer" in a culture that seems to require us to name ourselves *something*. If we consider, for a moment, that to refuse any sexual identity label at all might be a queer endeavor by resisting the (homo)normative push to name ourselves, what might we make of Brown's (2000) argument that silence reifies heteronormative power? Can our silence through a refusal to name resist a normative classification of our desires based on a binary system of gender? And/or might a reworking of identity labels be an act of resistance *through* naming?

(Re)Defining "Lesbian" and Refusing Labels

Cis partners expressed a fair amount of confusion over the language they want to use to describe their own sexualities, which was especially true for those who have lesbian-identified histories. While some are struggling with using "lesbian" and have switched to different sexual identity labels, others are resisting what "lesbian" means and are redefining the word in order to justify claiming it when in a relationship with someone who does not identify as female or "woman". Kate explained in a blog post:

> I suppose one of the main issues we have is that I identify as lesbian, which sort of clashes with his identity as male. Not that it bothers him, it is more how other people see us as a couple which is sometimes frustrating for me. I am proud of who I am, and I won't change it to suit anyone else's narrow definitions of sexuality.

As she illustrates, how she defines her own identity might be discordant with the ways that other people view her and her partner as a couple, but being read as straight doesn't mesh with her own identification as a lesbian. However, Kate is determined to claim "lesbian" even though she knows that it's contested – a move that could, perhaps, be considered a move to queer the label and resist who is "allowed" to claim it. She went on to write:

> I've been told numerous times I "must" be bisexual. I don't have a problem with bisexuals, their point of view makes a lot of sense to me, but I just don't

feel that I am one. The bottom line is, I would never have a relationship with a non-trans man and I'm still strongly attracted to women. There isn't really a word for that, so lesbian fits best out of the terms people recognize.

While some other participants took issue with "bisexual" as a limiting term suggesting only two genders were available, Kate doesn't like it for herself because it includes cis men. A lack of language for Kate to adequately describe her attractions has forced her to expand and redefine the currently available categories to fit. Joslin-Roher and Wheeler's (2009) participants had similar issues. They say:

> The failure of language to adequately describe nonheternormative desires and expressions may create challenges for lesbian and bisexual (and perhaps queer) identified partners of transmen in conceptualizing and verbalizing their identities and desires during and after transition.
> (Joslin-Roher and Wheeler 2009: 35)

Although I've suggested that resisting all categories might be a queer endeavor, it seems equally plausible that reworking the categories themselves could also be queer identity work, even if that identity is not called "queer". Renee, who claims a queer politics in relation to her lesbian identity, explains why there is no need to change her identity just because she's dating someone who identifies as male:

> At no point in time did I ever say to myself or think anything other than "I identify as a lesbian. I'm a lesbian." And I have my personal reasons for that, I have somewhat political reasons for that, there are many reasons why regardless of who I'm with I'm gonna identify as a lesbian. And I kind of just equated that with like, if you're bisexual and you happen to be dating a man you're not going to change your identity to straight just 'cause you happen to be dating a man.

Renee's political connections to "lesbian" – both in terms of a sexual identity and a larger community of women with whom she feels most comfortable – allow her to justify continuing to claim the label for herself. While she recognizes that her relationship isn't a lesbian relationship, she resists the notion that she should shift her own identity to be more (hetero)normatively in line with her partner.

However, claiming the identity of "lesbian" while being partnered with someone on the FTM spectrum is not without critique from other people, including other partners. For example, Sarah wrote to me in an email:

> I'm sorry, but if you're dating and in love with and attracted a guy (whose package, body, hair growth, smell, face, voice) has changed how can you call yourself a lesbian? Isn't that undermining your partner a little? It's like a girl who calls herself straight while dating a woman, it just doesn't make sense to me.

Sarah critiques how some cisgender women who are dating trans men use "lesbian" because it would be disrespectful to their partner. Whitley (2013) found in his work that,

> Common among significant others was the recognition that their perceived sexual orientation was relationally connected to their partner's transgender status. This meant that, regardless of how a significant other identified, when their partner chose to transition, their sexual orientation was put into question.
>
> (608)

Sarah's argument to pick a new category that affirms a trans partner's identity would not resonate with Kate and Renee, who argue that their sexuality does not change in response to a partner's gender identity, and who would likely resist policing around their identities. However, as Sanger points out:

> Individuals who are in intimate partnerships tend to find their sexualities defined, by themselves and/or others, with respect to the gender(s) of their partner(s). As a result sexual orientation identities must be theorized as relational.
>
> (2010: 101)

For Kate and Renee, sexual identity might be relational, but the relationality for them is one of resisting expected and/or normative categorizing.

However, while some partners were adamant that their identities did not shift in relation to having a trans partner (they instead redefined what those identities meant), others did experience a shift in identity, and/or language around identity, once they began dating a trans person or after their partner told them they were trans. This is not to say that a "new" identity has necessarily been solidified, but that a partner's transition sparked a shift or a questioning in some way; for some, this meant questioning the use of any identity label at all. Leah says in her vlog:

> I still label myself as lesbian but I'm not much for labels anyway because I don't feel that people, based on who they love or their sexual orientation, should have to put a label on anything because, you know, you should be able love who you want and it shouldn't be a big deal.

Leah still calls herself a lesbian, but also suggests that a move away from all labels might be desirable. Her statement that "it shouldn't be a big deal" points out the social importance of naming our sexual identities in intelligible ways. In a vlog, Sienna explains her relationship with labels and visibility:

> I think that labels are very dangerous things in the first place and I don't like to label my sexuality anymore, but as far as losing queer visibility

I have lost some of it because I am dating a trans man and he is a man and I don't identify as a lesbian anymore.

Sienna's visibility as a queer person was directly tied to her identification as a lesbian, as we also saw for other partners earlier, which she has mostly given up due to her partner being a trans man. While Renee and Kate challenge who can claim the label of "lesbian" by opening up the word to more possibility around gender, Sienna feels that if she is dating a man, then she cannot claim "lesbian" for herself (in line with Sarah's argument) and therefore wishes to not use labels. Interestingly, although Sienna makes it clear that her partner is a man, she does not indicate that she seeks to claim a straight identity. In her vlog, Beth presents some of the complexities of feeling that she can't use "bisexual" to describe herself because she's not attracted to either cisgender men or women:

> When I came out originally, I first came out as being bisexual years ago. And then I started identifying as lesbian and then I was identifying as queer or pansexual and now I just don't identify at all. . . . What I mean by "don't identify at all" is I don't subscribe to labels right now because I don't think I'm straight and I don't think I'm gay and I don't think I'm bi 'cause I'm not really attracted to women – I know I'm not attracted to women. And I'm not attracted to cisgendered [sic] men, at least not most of them. I'm mostly attracted to trans guys, but when you tell people that you're mostly attracted to trans guys then they call you a tranny chaser.

Beth notes that a failure of available terms to describe her attractions means that if she tells someone she is primarily attracted to trans guys, she is labeled a "tranny chaser" – a label with negative connotations with which she does not identify.[10] Lacking language adequately describes her attractions and sexuality, and a danger around speaking these attractions at all, has led Beth to simply not identify with any sexual identity label. Using labels has become unimportant for Reagan as well, based on who she's dating and the complications around language in describing her sexual identity:

> I'm still totally attracted to women, I'm dating a boy – I dunno! What do you call that?! I dunno. Maybe this sounds dumb and contradictory when I did work so hard to find this sense of self from coming out and being gay to being able to say now that that's not that important to me anymore. Maybe dating a trans guy is just putting those things into perspective for me and realizing that, you know, maybe it's not that important what people think.

What is particularly interesting about what Reagan says in her vlog is that "dating a trans guy is just putting those things into perspective". That is, Reagan has rethought identity labels and their usefulness overall, not just in the context of her current relationship with a trans person. Further, she questions the importance of coming out since the available identity labels cannot capture who she is or who she is dating.

Other partners are questioning their sexuality, but haven't decided to give up on identity labels like Leah, Sienna, Beth and Reagan have. Autumn articulates the complexities of her own identifications in concert with her relationship in one weekly vlog:

> Of course, I identify, or identified, I don't even know at this point, as a lesbian. With that being said, it made me question who I was because [my partner] has always been a man. . . . So, that's what led me to question my sexual orientation and how I identify. It made me think, "Well, if I have the capacity to love a man for seven years, am I really gay? What does that make me? Does it make me bisexual? Does that make me pansexual? Does it make me a lesbian who just so happened to fall in love with a trans guy?" I don't know what me being in love with [my partner] makes me. And I'm not sure if I have to know that, but I don't know that.

While Autumn is questioning how her relationship might change her sexual identity, she's also grappling with the fact that she is supposed to know what her sexual identity is. That is, Autumn recognizes the larger social discourse around the need to know oneself and to be able to articulate that for others (see Butler 2005; Foucault 1978, 1984; Wilchins 2004).

While straight-identified partners have continued to either identify as straight or shift their identity to "queer" or "pansexual" since being in a relationship with someone on the FTM spectrum, none of the partners who identified as "lesbian", "bisexual", "pansexual" or "queer" *before* being with a trans person shifted their identity to straight. This latter group of partners either shifted their identity to another category that seemed more open, refused sexual identity categories altogether or redefined the categories in ways that include their experience and relationship. While some of the partners were willing to change how they referred to their relationship by saying it was a straight relationship, they were not willing to adopt a sexual identity of "straight". This suggests that the personal and political connections that cis partners have to "lesbian", "bisexual", "pansexual" or "queer" might hold more weight than a trans partner's gender identity when contemplating a shift in sexual identity. While Ward (2010) argues in her work that femme partners of people on the FTM spectrum engaged in "gender labor" to affirm their trans partners' gender identities, the narratives in my project suggest that some cis women actively refuse to do that labor in relation to their own sexual identities, even if they may redefine the relationship itself.

Queer(ing) Self

Some participants drew on queer political and activist language to define their identities as being open, fluid and shifting, or to resist identity labels that require a binary notion of gender for intelligibility (see Butler 2004; Gamson

1996; Warner 1999). In contrast to the previous discussion of the challenges regarding redefining or resisting labels for sexual identity, some participants have actively embraced the complexities of language when it comes to their relationship with a trans person by adopting explanations of queer selfhood. As Dakota explained at length in an email to me:

> To me, identifying as queer is a way to say that my desires and attractions (emotional, sexual, physical, etc.) are non-normative, that I am interested in a wide variety of people with a diverse range of gender identities and expressions. I don't fit into a straight or lesbian label, and I find "bisexual" problematic as it upholds a gender binary, and because the term has so many negative connotations. . . . I also find that identifying as queer gives me common ground with gay men, trans women, stone butches, high femmes, and others with whom I might not seem to have a lot in common. To me, "queer" is also a reclaimed identity (although it originally did come out of the queer community itself, it was often used in a pejorative way) and an intentionally politicized one.

Dakota's explanation of her identity utilizes some of the similar frameworks as previous explanations I've presented, but is quite different in content. She mentions the identity labels that she's not comfortable with, similar to other participants, and why she's not comfortable with them; but, instead of settling on something that simply seems like it fits or continuing to be confused about where she fits, she claims a queer location that is as much about her relationship and desires as it is about her queer politics. That is, for Dakota, there is little separation of identity from the political importance of finding common ground with others that share similar politics and non-normative senses of self. This is in line with Michael Warner's (1999) discussion about "queer" being more than about sexuality and identity – it operates as a specific cultural politics of resistance. Warner argues that being in queer culture "is a way of transforming oneself, and at the same time helping to elaborate a commonly accessible world" (1999: 71). In other words, Dakota's queer politics are not a politics of identity; they are a politics of encouraging a shift from the normative that allows and celebrates coalitions around difference.

In a vlog, Chloe also explained her identity in relation to queer (and lesbian) politics:

> I've really shied away from using the word lesbian. I only use it in particular situations for particular connotation. I think that for me, lesbian feminism has a very specific history that I like, so when I'm talking about myself as a feminist I think about myself as a radical queer feminist or a progressive feminist, but there's a lot of lesbian feminist history that I really align with and feel that I'm a product of. So, I'll use the word lesbian in feminist spaces to denote a certain genealogy that I think I'm coming from with my activism and my education stuff. . . . But almost all the time – 95% of the time

probably now – I identify as queer. I say I'm queer-identified because I like the fact that it confuses people. I don't like it and I like it. I like the fact that it confuses people because people are like, "What does that mean?" and then that opens up conversation about how desire, orientation, identity, all those things are far more complicated than binaries allow them to be.

Chloe does use the "fits me best" language of identity that we've heard before; however, she provides reasons for her historical connections to "lesbian" as related to a lesbian feminism that she feels she is "a product of". For Chloe, challenging and questioning are key parts to her explanations of a queer self, which follow Butler (1990, 1991, 1993) and others (see Gamson 1996; Warner 1999) given her use of "trouble making" in relation to identity politics. What is notable here is that both Dakota and Chloe have fashioned queer identities that place importance on notions of questioning, challenging or transgressing, and comfort with fluidity and complexity. This is in contrast to previous discussions of identity that I presented where partners talked about being uncomfortable with being perceived in ways that they did not identify, held strongly onto concrete identity labels, challenged ways that others were using those labels and/or had just decided to give up on identity labels completely. Instead, here we see partners very intentionally claiming political identities that they accept may confuse others. That said, some partners discussed their queerness simply as an alternative to other categories (see Baker 2008). As Sarah noted in an email:

I've never felt comfortable with the lesbian label, I like/am attracted to gender ambiguity, and I don't look like someone from the *L Word* (which is what I picture when I think "lesbian") but I'm definitely not straight, as I like women, so queer seems more fluid and relaxed and open, and I like that freedom.

Meghan also considers "queer" a more open identity than others she has come across:

I now choose to use "queer" to describe my sexual orientation because "bisexual" isn't a good fit anymore. I like to think of queer as being a broad term, encompassing more than just two genders. I like to think of queer as somewhat of a label for people who don't like labels.

For this group of partners, "queer" operates as an identity that is open, fluid, politically engaged and/or connected to their partners' transness through needing an identity that also includes trans people. That is, for these partners, "lesbian" and "bisexual" don't work because they are too limiting to describe their attractions and their relationship with their trans partner and they aren't seeking to redefine those words to fit their relationships, attractions or desires. While redefining "lesbian" and "bisexual" may work for some, as illustrated previously, Dakota, Chloe and Meghan recognize that these words have different (and limited) meaning to others, and they often don't include trans people.

For this group of cis partners, "queer" is often used to resist normative politics around gender and sexual identities in larger society and in "lesbian and gay" contexts (Warner 1999). "Queer" is explained as a reclaimed and antinormative identity by these cis women, where trans partners might be included through the language used to validate the cis women's identities, despite not referring to transness specifically.[11]

Conclusions

The problematics of identity in cis/trans relationships can be connected to Judith Butler's position around the speaking of identity and coming out. As she posits,

> If I claim to be a lesbian, I 'come out' only to produce a new and different 'closet.' The 'you' to whom I come out now has access to a different region of opacity. Indeed the locus of opacity has simply shifted: before, you did not know whether I 'am,' but now you do not know what that means.
>
> (Butler 1991: 15–16)

For the white, middle-class, cis women partners whose stories are included here, this means that their coming out as being "straight", "bisexual", "lesbian" "queer" or "pansexual" doesn't tell the whole story, fails to capture the gender identities of both partners in the relationship and the use of these terms may be contested by others.[12] In short, these labels fail to refer to transness and none of the available sexual identity categories we have in contemporary US culture are able to describe the complexities of gender and desire in cis/trans relationships. That is, almost any sexual identity that is claimed automatically erases the trans specificities of the relationship. As Hines says, "Dominant categorical frameworks are unable to account for the complexities of gendered and sexual transformations" (2010: 150). For some, this is desirable because they identify as straight and their trans partner prefers not to be read as trans. But, for others, outness and visibility is a key part of their sexual identity. This is in line with Joslin-Roher and Wheeler's (2009) research, where their participants also spoke about a need for identity language that includes trans people. While moving beyond a focus on identity may seem desirable, naming oneself with recognizable terms allows us to find similar others with which to fashion community. As I've noted elsewhere (Tompkins 2011, forthcoming), lacking language to describe a sense of self can often prevent cis people with trans partners from forming communities with others who share their experience.

Notes

1 I use "trans" in the broadest sense throughout this chapter, referring to any and all identities that are not cis identities. With regard to the data for this specific research project, "trans partner" and "FTM spectrum" refer to people who were assigned female at birth who do not identify as female and/or woman (e.g., FTM, genderqueer, trans man, pangender, non-binary, gender non-conforming). I realize that these terms may not work

for everyone, though the intention is inclusivity even if these terms may have been used in exclusive ways elsewhere.

2 This is evidenced in part by the number of US organizations that are based on LGBT(Q) identity and involved in national politics, such as the Human Rights Campaign, National Gay and Lesbian Task Force, National Center for Transgender Equality, Gay and Lesbian Alliance Against Defamation, and the Gay Liberation Front, to name a few.

3 This particular research focused only on the experiences of cis women in the relationships; I did not interview their trans partners. Likewise, there are no trans-identified people from the YouTube channels as the vlogs were also focused on cis experiences only. See Hines (2007) and Sanger (2010) for research that includes trans narratives about intimate relationships. Certainly, research on other gender configurations in a partnership (e.g., cis man and FTM, cis man and MTF) may yield different questions and results.

4 I did not ask about the specific identities of their trans partners as the trans partner was not the focus of this project, though some participants mentioned identities during the interviews (e.g., trans guy or FTM).

5 See Tompkins (2011) for the full research project.

6 "Vlog" is short for "video blog" and usually refers to videos made by people to discuss their personal lives.

7 The cis partners who made the videos I used as data all reside in the US and speak to issues that may be specific to the US as a whole or to specific areas/communities within the US.

8 Most names are pseudonyms either chosen by participants or given by me. Five of the 18 interview participants chose to forgo confidentiality by requesting to use their legal names in all work produced from this project. Pseudonyms were given to all who created YouTube vlogs.

9 I'm not suggesting here that all cisgender partners wish for queer visibility and all trans people do not, but it is these cases where issues of visibility are most contested within the relationship and seem most pressing for cis partners.

10 For a more in-depth discussion about "tranny chaser" and Beth's narrative, see Tompkins (2014, forthcoming).

11 While one could argue that this is a form of gender labor similar to what Ward discusses in her work, of notable difference is that the cis women in my project are focused on themselves instead of "giving gender to others" (2010: 237).

12 These categories/labels can also be problematic for trans people in their relationships (see Hines 2007; Sanger 2010, 2013).

References

Baker, Paul. 2008. *Sexed Texts: Language, Gender, and Sexuality*. Oakville, CT: DBBC

Boufoy-Bastick, B. 2004. "Auto-interviewing, Auto-ethnography and Critical Incident Methodology for Eliciting Self-conceptualised Worldview." *Forum Qualitative Sozialforschung / Forum: Qualitative Social Research* 5(1). Retrieved from www.qualitativeresearch.net/index.php/fqs/article/view/651/1410

Brown, Michael P. 2000. *Closet Space: Geographies of Metaphor from the Body to the Globe*. New York: Routledge.

Brown, Nicola R. 2009. " 'I'm in Transition Too': Sexual Identity Renegotiation in Sexual Minority Women's Relationships with Transsexual Men." *International Journal of Sexual Health* 21(1):61–77.

Brown, Nicola R. 2010. "The Sexual Relationships of Sexual-Minority Women Partnered with Trans Men: A Qualitative Study." *Archives of Sexual Behavior* 39:561–572.

Brown, Nicola R. 2005. *Queer Women Partners of Female-to-Male Transsexuals: Renegotiating Self in Relationship*. PhD dissertation, Department of Psychology, York University. Retrieved from ProQuest Dissertations & Theses Database, 1079667931.

Butler, Judith. 1990. *Gender Trouble: Feminism and the Subversion of Identity*. New York: Routledge.

Butler, Judith. 1991. "Imitation and Gender Insubordination." In Diana Fuss (ed), *Inside/ Out*. New York: Routledge. Pp. 13–29.

Butler, Judith. 1993. *Bodies That Matter: On the Discursive Limits of "Sex"*. New York: Routledge.

Butler, Judith. 2004. *Undoing Gender*. New York: Routledge.

Butler, Judith. 2005. *Giving an Account of Oneself*. New York: Fordham University Press.

Cromwell, Jason. 2006. "Queering the Binaries: Transsituated Identities, Bodies, and Sexualities. In Susan Stryker and Stephen Whittle (eds), *The Transgender Studies Reader*. New York: Routledge. Pp. 509–520.

Dicks, Bella, Bambo Soyinka, and Amanda Coffey. 2006. "Multimodal Ethnography." *Qualitative Research* 6(1):77–96.

Fontana, Andrea. 2003. "Postmodern Trends in Interviewing." In J.F. Gubrium and J.A. Holstein (eds), *Postmodern Interviewing*. Thousand Oaks, CA: Sage. Pp. 51–65.

Foucault, Michel. 1978. *The History of Sexuality: An Introduction, Vol. 1*. New York: Vintage.

Foucault, Michel. 1984. "The Subject and Power." In H. Dreyfus and P. Rabinow (eds), *Michel Foucault: Beyond Structuralism and Hermeneutics*. Chicago: University of Chicago Press. Pp. 208–226.

Gamson, Joshua. 2000. "Sexualities, Queer Theory, and Qualitative Research." In Norman K. Denzin and Yvonna S. Lincoln (eds), *Handbook of Qualitative Research* (2nd ed). Thousand Oaks, CA: Sage. Pp. 347–365.

Gamson, Joshua. 1996. "Must Identity Movements Self-Destruct? A Queer Dilemma." In S. Seidman (ed), *Queer Theory/Sociology*. Cambridge, MA: Blackwell. Pp. 395–420.

Green, Eli R. 2006. "Debating Trans Inclusion in the Feminist Movement: A Trans-Positive Analysis." *Journal of Lesbian Studies* 10:231–248.

Hines, Sally. 2007. *TransForming gender: Transgender Practices of Identity, Intimacy and Care*. Bristol, UK: Policy Press.

Hines, Sally. 2010. "Sexing Gender; Gendering Sex: Towards an Intersectional Analysis of Transgender." In Y. Taylor, et al. (eds), *Theorizing Intersectionality and Sexuality*. New York: Palgrave Macmillan. Pp. 140–162.

Hookway, Nicholas. 2008. "'Entering the Blogosphere': Some Strategies for Using Blogs in Social Research." *Qualitative Research* 8(1):91–113.

Joslin-Roher, E., & Wheeler, D. P. 2009. Partners in transition: The transition experience of lesbian, bisexual, and queer identified partners of transgender men. *Journal of Gay & Lesbian Social Services, 21*, 30–48.

Mason, Michelle E. 2006. *The Experience of Transition for Lesbian Partners of Female-to-Male Transsexuals*. PhD dissertation, The California School of Professional Psychology, Alliant International University. Retrieved from ProQuest Dissertations & Theses Database, 1394648401.

Murthy, Dhiraj. 2008. "Digital Ethnography: An Examination of the Use of New Technologies for Social Research." *Sociology* 42(5):837–855.

Nyamora, Cory M. 2004. *Femme Lesbian Identity Development and the Impact of Partnering with Female-to-Male Transsexuals*. PsyD dissertation, The California School of Professional Psychology, Alliant International University. Retrieved from ProQuest Dissertations & Theses Database, 766113091.

Pfeffer, Carla A. 2009. *Trans(Formative) Relationships: What We Learn About Identities, Bodies, Work, and Families from Women Partners of Trans Men*. PhD dissertation, Department of Sociology, University of Michigan. Retrieved from Deep Blue at the University of Michigan, http://hdl.handle.net/2027.42/63628.

Pfeffer, Carla A. 2010. "'Women's Work'? Women Partners of Transgender Men Doing Housework and Emotion Work." *Journal of Marriage and Family* 72(1):165–183.

Pfeffer, Carla A. 2017. *Queering Families: The Postmodern Partnerships of Cisgender Women and Transgender Men.* New York: Oxford University Press.

Phelan, Shane. 2004. "Alliances and Coalitions: Nonidentity Politics." In D. Carlin and J. DiGrazia (eds), *Queer Cultures.* Upper Saddle River, NJ: Pearson Education, Inc. Pp. 700–719.

Richardson, Laurel. 1988. "The Collective Story: Postmodernism and the Writing of Sociology." *Sociological Focus* 21(3):199–208.

Rubin, Gayle. 2006. "Of Calamities and Kings: Reflections on Butch, Gender, and Boundaries." In Susan Stryker and Stephen Whittle (eds), *The Transgender Studies Reader.* New York: Routledge. Pp. 471–481.

Sanger, Tam. 2010. *Trans People's Partnerships: Towards an Ethics of Intimacy.* New York: Palgrave Macmillan.

Sanger, Tam. 2013. "Trans People's Partnerships: Rethinking the Limits of Relating." In T. Sanger and Y. Taylor (eds), *Mapping Intimacies.* New York: Palgrave Macmillan. Pp. 171–189.

Sedgwick, Eve Kosofsky. 1990. *Epistemology of the Closet.* Berkeley: University of California Press.

Tompkins, Avery Brooks. 2011. *Intimate Allies: Identity, Community, and Everyday Activism Among Cisgender People with Trans-Identified Partners.* Doctoral dissertation. Retrieved from SUrface at Syracuse University, http://surface.syr.edu/soc_etd/67.

Tompkins, Avery Brooks. 2014. "'There's No Chasing Involved': Cis/Trans Relationships, 'Tranny Chasers,' and the Future of a Sex-Positive Trans Politics." *Journal of Homosexuality* 61:766–780.

Tompkins, Avery. Forthcoming. "Trans Sexualities." In Nancy L. Fischer and Laurel Westbrook (eds), *Introducing the New Sexuality Studies: Original Essays* (4th ed). New York: Routledge.

Valentine, David. 2007. *Imagining Transgender: An Ethnography of a Category.* Durham, NC: Duke University Press.

Valocchi, Stephen. 2005. "Not Yet Queer Enough: The Lessons of Queer Theory for the Sociology of Gender and Sexuality." *Gender and Society* 19(6):750–770.

Vidal-Ortiz, Salvador. 2002. "Queering Sexuality and Doing Gender: Transgender Men's Identification with Gender and Sexuality." *Gendered Sexualities* 6:181–233.

Ward, Jane. 2010. "Gender Labor: Transmen, Femmes, and Collective Work of Transgression." *Sexualities* 13(2):236–254.

Warner, Michael. 1999. *The Trouble with Normal: Sex, Politics, and the Ethics of Queer Life.* New York: The Free Press.

Whitley, Cameron T. 2013. "Trans-Kin Undoing and Redoing Gender: Negotiating Relational Identity Among Friends and Family of Transgender Persons." *Sociological Perspectives* 56(4):597–621.

Wilchins, Riki Anne. 1997. *Read My Lips: Sexual Subversion and the End of Gender.* Ithaca, NY: Firebrand Books.

Wilchins, Riki. 2004. *Queer Theory, Gender Theory: An Instant Primer.* Los Angeles, CA: Alyson Books.

12 "When a Trans Is Killed, Another Thousand Rise!"

Transnecropolitics and Resistance in Brazil

Joseli Maria Silva, Marcio Jose Ornat, Vinícius Cabral and Débora Lee Comasseto Machado

Introduction

In meetings of political training and activism for *travestis*, transgender and transsexual people,[1] – with their fists to the air – repeat many times the sentence that is the title of this chapter "when a trans is killed, another thousand rise". This catch phrase is followed by the names of murder victims and creates a kind of mourning ritual. Pain is expressed publicly for the loss of lives that ended tragically. The ritual highlights the revolt and willingness to fight against the violence experienced on a daily basis in the Brazilian transphobic society.

Travestis and transsexual women belong to a group that is highly vulnerable to violence and early death in Brazil. Despite the lack of systematic studies on the life expectancy of female transsexuals and *travestis*, Antunes (2013) states that the life expectancy of this group is around 35 years, while the Brazilian population tends to live up to 74.9 years (IBGE 2013). The non-governmental organization Transgender Europe, which monitors murders of transpeople around the world, shows that Brazil is the country with the highest number of reported hate crimes (Balzer, Hutta, Adrián, Hyldal and Stryker 2012). The Brazilian non-governmental organization Grupo Gay da Bahia (2014) note in their 2014 annual report a total of 326 murders of people belonging to the LGBTI community in Brazil, and out of this number, 134 were *travestis*. The Associação Nacional de *Travestis* e Transexuais – ANTRA (National Association of *Travestis* and Transsexuals) – reported 117 deaths of transpeople in Brazil between January and August 2017.[2]

Violent deaths that victimize *travestis* and transsexuals in Brazil are the result of complex relations and practices of material and symbolic violence that cross simultaneously the social, economic and cultural structures and reach the bodies that transgress compulsory heteronormativity (Butler 1993). *Travestis* and transsexuals are exposed to death every day in a continuous process of dehumanization of their precarious lives (Butler 2004).

Although death is an everyday and universal occurrence, the experience of death is mediated by the intersections of body, culture, society and state (Maddrell and Sidaway 2010). Therefore, the death of the body is not only a natural phenomenon, it also has socially, temporally and spatially meaning.

DOI: 10.4324/9781315613703-12

This chapter analyzes how *travestis* and female transsexuals understand the relationship between death, place and space, living under the logic of transnecropolitics existing in Brazil. The transnecropolitics idea is inspired by Mbembe (2003 and 2008), who discusses death management in contemporary societies. In order to maintain a "good society", it is deemed necessary to eliminate those beings who are socially considered as enemies. Thus, we understand that the death of *travestis* and female transsexuals in Brazil is the result of a state policy that kills or allows the death of people who are considered a social hazard or whose lives are considered socially useless. Twelve people took part in the research; from these, six self-identified as *travestis* and the other six as female transsexuals.[3]

The interviewees were stimulated by reading about and viewing images of six news items presented in online papers about the death of transpeople. After reading the news, they developed their narratives based on two investigative axes. The first comprised their own opinion about the piece of news, while the second focused on their perspective in relation to their own death. The narratives were analyzed based on the content analysis by Bardin (2004), which enabled the construction of a semantic network of meanings involving the death of transpeople.

The chapter is organized in two parts. In the first section we argue that the transpeople who are murder victims in Brazil illustrate intersections of sexuality and class in which, by defying the biopolitical calculation logic, they are considered as superfluous beings by the Brazilian society. The second part of the chapter highlights the different understandings created by *travestis* and female transsexuals about their precarious conditions and exposure to death.

"May God Protect Me, Because I'm Going, but I Don't Know If I'm Coming Back": Trans Abandonment and Death as an Order Maintenance Politics

Risk and exposure to death are everyday elements in the life of Brazilian *travestis* and trans women, mainly those that live by sexual commercial activity. Azaleia reports this perception by saying

> as I have lived in many houses, with many prostitutes, we have this prayer that we say before going out to work. 'May God protect me, because I'm going, but I don't know if I'm coming back'. We are faithful, but we know that we are in constant danger and we try not to think about that and just live.
>
> (Interview carried out with Azaleia, 25/08/2013, in Curitiba)

Prostitution is the activity from which 90% of the *travestis* and trans women engage in to survive economically in Brazil, according to ANTRA. The prostitution territory is simultaneously a place of life and death to the group. *Travestis* and trans women say that the prostitution territory is a space where their femininity is recognized and desired.[4] Moreover, it is the space where they develop friendships, solidarity and protection (Ornat 2009). However, this area is also a space of high risk to murder. Azaleia explains that during "the working time, when we need to be on the streets, seeking to make a living, our survival is dependent on working where we can also be raped and murdered". (interview with Azaleia, on 25/08/2013, in Curitiba).

Coexisting with vulnerability to death is a reality that transpeople have to deal with on a regular basis, since there are no specific public policies or laws to protect this group in Brazil. The mortality and death statistics have been produced by entities and non-governmental organizations which take part in the LGBTI movement and are based on the news spread on the Internet by blogs, papers and sites.[5]

The Brazilian Secretariat of Human Rights has a telephone line for reporting violence and protection against the violation of the human rights of minority groups (afro-descendants, homeless, disabled and LGBTI people). This project is called "Disque 100" and was initially supported by non-governmental organizations and focused on fighting children and adolescents abuse. In 2003, this service was institutionalized, and the federal government became responsible for it and in 2010, the scope of the project was extended to include other vulnerable groups. It seems relevant to emphasize that it was only in 2012 that official data were disclosed regarding the violation of the human rights of the LGBTI population, based on that channel of communication.

In the data originated from the "Disque 100", the transpeople appear as an invisible group regarding violence reports. Only 1.47% of the total number were related to violence committed against *travestis* and 0.49% against transsexual people (Brasil 2013, p. 24). Data presented by the state show inconsistency, indicating that the cases of violence and the number of deaths among transpeople are even higher than that reported.

ANTRA carried out an important task in 2017 and calculated the murders of *travestis* and transsexuals in Brazil. From the total of 115 murders registered by the institution in 2017, between January and August, 5.21% were trans men, 10.43% were trans women and 84.34% were *travestis*. This shows that there are internal differences in the group of transpeople who are victims of murder. There is a pressing need to broaden the understanding of the elements that aggregate different levels of vulnerability to the violent death of the trans population in Brazil.[6]

The average life span of transpeople victims of murder is 27.9 years. Antunes (2013) points out that the life span of the trans population in Brazil is 35 years and the *Transrevolução*[7] group reports a life span of 30 years. These estimations of the life span of the transpeople consider other factors in addition to the lethal violence such as death due to HIV- and AIDS-related illnesses, plus complications and procedures of body transformation[8] without suitable medical care.

The means used to kill the transpeople are varied. Less than 2% of the murders were through stoning (1.7%), car crash (1.7%), strangling (1.7%), hit with wood (1.7%), burning (0.9%) and suffocation (0.9%). There is no information available about the remaining cases (1.7%). Shooting was the main means used in 51.3% of the murders, 18.3% deaths resulted from stabbing, 12.2% from beating and 7.8% of the murders resulted from more than one violent attack, involving cruelty, torture and public humiliation of the victims' suffering.

Cabral, Silva and Ornat (2013) point out that the space is of crucial importance for the murderers' method of killing. In places where there is higher flow of people, the victim's execution is fast, usually involving a shotgun. However, when the murderer manages to meet the victim in places of fewer or no people, the victim is executed with barbarism, psychological violence and physical torture. In such conditions, the main victims are *travestis* that have their bodies mutilated, objects penetrated into their anus, are dragged throughout the streets, thrown from overpasses and shot.

The group most targeted by murderers are young, poor *travestis*, who, in their majority, make a living working in the sex industry. It is the intersectionality (Crenshaw 1991; Collins 2000; McCall 2005; Davis 2009) of these elements (age, gender, sexuality, occupation) that increase the chances of death, evidencing, according to Puar (2007), that there are particular ways of social hierarchy of the lives that matter socially, even within the already stigmatized LGBTI population.

The bodies of young *travestis* are quintessentially political objects, since

> the body implies mortality, vulnerability, agency: the skin and the flesh expose us to the gaze of others, but also to touch, and to violence, and the bodies put us at risk to becoming agency and instrument of all these as well.
> (Butler 2004, p. 26)

They are people whose bodies are the target of disciplinary intervention, as pointed out by Foucault (1978), which starts at the earliest in the person's life and is made spatially viable through the everyday experience of the gender tyranny (Doan 2010). *Travestis* and transsexual people face rejection in their own homes and also at school, hampering their chances to obtain better work and income opportunities in their adulthood (Boulevard 2013; Lee 2013; Nikaratty 2013). ANTRA claims that 57% of the transpeople have not even concluded elementary school and just 0.02% managed, only recently, to do higher education courses.[9]

There are structural forces that gradually build the path to "slow death" (Berlant 2007), and violence for *travestis* and transsexual women. The end of the journey peaks with their extermination in brutal ways, since, according to Foucault (1995), torment is part of the social ritual of punishment. Martyrdom must be ostentatious in order to become an example of social order, making the victim look despicable and in addition to the lethal aggression, symbolic practices of humiliation must be performed.

The material existence of poor people, who do not conform to the gender assigned at birth, is a type of rebellion against the biopolitical apparatus in Brazil (Foucault 1997). The life of *travestis* and transsexual women results from a set of relations between macropolitical power structures and micropolitical techniques through which their bodies are interwoven with nations, states and capitals (Stryker, Currah and Moore 2008). Their lives are not considered viable, and they become subjects living a thin border between life and death (Foucault 1997).

The establishment of boundaries between those who have their lives secured and those who are directed toward death is based on hierarchical systems of values around economic, cultural and social differences that are built up through discourse as something natural. Such hierarchy system has defined the lives of *travestis* and transsexual women as something derogatory, as if they were socially undesirable beings (Stryker 2014).

The deaths of *travestis* and transsexual women can be categorized as an ordered logic of power representation and the way their bodies and lives are understood as socially disposable. The link between life and death of transpeople can only be understood through social, cultural and economic policies that create mechanisms that result in trans death (Agamben 1998).

Life and death are political issues, whose management depends on the relations of power and hierarchical valuation of lives that must be preserved or wasted. Therefore, the death of transpeople occurs, as argued by Agamben (1998) and Mbembe (2003), as part of work, as a practice, and ultimately, as an outcome the unequal distribution of power, reaching lives classified as disposable.

"One Less to Disturb": Transnecropolitics Narratives

Narratives developed by people who took part in the research were systematized through the content analysis that enabled the organization of a semantic network comprising three distinct discourses. The first is around the category of intersectionality, highlighting the group awareness of the different levels of vulnerability to murder among transpeople. The second reveals their understanding of social and economic processes that constitute the trajectory of death of *travestis* and transsexual women. The third verified is the awareness of the creation of a social enemy and the consequent trivialization of the death of *travestis* and transsexual women in prostitution spaces.

Chart 12.1 shows the general characteristics of the group that took part in the research.

The *travestis*' and transsexual women's discourses show full knowledge of the specific conditions that support unequal regimes of life and death, having specific intersections between poverty, gender and sexualities (Haritaworn, Kuntsman and Posocco 2014). The excerpt from Rosa's testimony is a good example in the sense of arguing that among the transpeople there are those who are more vulnerable to murder than others:

> I see that there are people who are more likely to be killed, because they face a higher level of risk. These are the *travestis*, sex professionals. I, for example, even being a transwoman, I suffer less prejudice than those who

Chart 12.1 General profile of the group taking part in the research

Fictitious name	Gender self-identification	Age	Self-definition of economic activity
Amarilis	Transwoman	23	Hairdresser/prostitution
Azaleia	Transwoman	40	Activist/prostitution
Cravina	Transwoman	19	Student/prostitution
Estrelícia	Transwoman	20	Student/prostitution
Girassol	Transwoman	21	Activist
Iris	travesti	42	Activist/prostitution
Jasmin	travesti	22	Prostitution
Lirio	travesti	21	Activist
Margarida	Transwoman	37	Hairdresser/prostitution
Rosa	travesti	50	Activist/cleaner
Tulipa	travesti	34	Prostitution
Violeta	travesti	25	Prostitution

are on the streets, exposed until dawn. I managed to get a job here at the NGO now and that helps me scape the risk of death. We are trans in the same way, but *travestis* are more exposed to being killed than I am, now that I have another occupation.

(Interview with Rosa on 07/09/2013, in Curitiba)

The interviewees pointed out that there is a social hierarchy among *travestis* and transsexual women in relation to the possibility of making an economic livelihood out of prostitution, which minimizes the risk of being murdered. The *travestis* and transwomen who do not have sex reassignment surgeries[10] are more vulnerable to murder. The alignment of bodies to the gender norms creates, according to them, higher probability of having other ways of making a living, other than prostitution, reducing the risks of being killed.

The narratives emphasize that there is the constitution of a transnormativity[11] that divides transpeople into two groups (Johnson 2016; LeBlanc 2010; McDonald 2006). "Respectable" transsexual women are those who change their genitalia to adapt to gender norms. "Others", who do not have gender surgeries, are considered transgressors. These are the ones who are most vulnerable to violence and murder, and they are mainly *travestis*.

The second discourse community that takes part in the semantic network resulting from the narratives of transpeople interviewed reveals the awareness of social and economic processes that establish their civil death and they question the reasons for the creation of a necropolitics around their lives. Margarida's testimony is one example of this political view:

I don't know how much a trans life is worth. I know that in Brazil we experience family and institutional violence. I have suffered both. Since the first moment you identify yourself as a trans, you learn you're wrong, that you're not worth anything, that you're ill. Your family repeat 'you must have some disorder'. If you listen to that coming from your dad or mom, who can tell you the opposite? Convince you that you're worth it? Everybody says the same thing. After the family, comes the school, the church, and so on. We aren't recognized, we don't have the right to a name, health or family. . . . Nobody asks how you ended up in this situation of prostitution and they think it is your fault when you're killed. They say 'good, one less to disturb' the society. Because you're marginal, criminal, dangerous. . . . I don't understand why people get surprised when they see *travestis* and transsexuals trying to study. I don't understand why we're denied this. I don't understand how much our lives are worth, but I'm a human being and I fight for people to see me the way I am, as a human being.

(Interview with Margarida, on 15/09/2013, in Curitiba)

The narratives show that *travestis* and female transsexuals resist and denounce the abandonment suffered in their existence; they are aware and fight against this dehumanization process. They recognize the necropolitics that creates the possibility of destruction of their lives, simply because they are considered disposable lives, deaths which are not mourned socially (Agamben 1998 and Mbembe 2003).

The third discourse community in the semantic network emphasizes how they understand the way the Brazilian society creates, through discourse, the idea of the "enemy", the one that must be destroyed so that the beings that really matter can live. The image of *travestis* and female transsexuals described by Iris reveals the awareness of this process, describing her social image: "a *travesti* is seen as an aberration, a monster and automatically people don't want to get closer, because they're afraid of us. Because we're seen as drugged, bandit, rubbish" (Interview with Iris, on 17/08/2013, in Ponta Grossa).

The transnecropolitics is exercised on a daily basis to mark the lives that are considered good and those that are bad (Butler 2004). The necropolitical calculation aims at protecting normative lives while sacrificing trans lives (Hutta 2013). The physical death of a group that is already socially dead is the predictable ending according to the trans narratives, as shown:

Being excluded in life I think is the worst thing. But there's a hygienization process in the society. For the society, it's extremely normal to see a *travesti* being killed. This is a usual image. Violence, violent and cruel death are always linked to what the society think of us. The prostitution and promiscuity stigma. I think that's why there is so much violence . . . then I believe the society applaud the deaths. You can put it there in your research. I say: the society applaud violence against *travestis* and transsexual people. I want you to emphasize this I'm telling you. Because that's the reality . . . the society applaud that other people do the dirty work that they would like to be done, that is to exterminate us, *travestis* and transsexual people.

(Interview with Margarida, on 15/09/2013, in Curitiba)

Margarida's report shows the group awareness that there are several ways of "making die" or "letting die" or "exposing some people to death" by creating situations of structural negligence in relation to *travestis* and female transsexuals. The Brazilian society creates and nourishes specific power governance that distribute at different levels violence, death and the weakening of the transpeople lives (Butler 2004).

The constant presence of the perspective of death experienced by Brazilian *travestis* and trans women represents a short-sighted life style. Few long-term life plans are made, which makes their vulnerability even more serious. They end up neglecting themselves, their health, pension plan and take risks in dangerous situations, since they see no future. Rosa says:

> I don't think or imagine myself getting old. Maybe, exactly because I think I'll never get there. Then, I live every day intensely, because I'm not going to grow old. . . . It seems that *travestis* and transsexuals are determined not to make it to the third age.
>
> (Interview with Rosa on 07/09/2013, in Curitiba)

The transnecropolitics is operated through the creation of boundaries, establishing places where the "enemy" can or cannot circulate in the town. The narratives point out various spaces of exclusion and the awareness that prostitution areas are spaces that at the same time make possible the life and death of *travesties* and transsexuals. The same space of prostitution that allows them economic support, and therefore, a life, is the space where they are most vulnerable to violence and death. This simultaneity of opposed meanings "life and death" is created from the starting point of an intentional space politics that makes transnecropolitics viable, as reported by Azaléia:

> The same hypocritical society that condemn and kill you, is the one that supports you in the prostitution. The street is the place of highest vulnerability, because you go out with somebody you don't know and never know whether you'll be back or not, because this person can be a good person, but they can also be bad. In prostitution there is a lot of dispute for power, money and drugs, which makes the vulnerability to violence and death much higher, I think.
>
> (Interview with Azaleia, on 25/08/2013, in Curitiba)

The transnecropolitics does not end with deaths; rather, it is kept alive through memory work of those that are deemed to be "enemies". *Travestis'* and transsexuals' narratives about suffering and death show the construction of invisibility of their female existence. Jasmim creates a narrative around the concealment of their trans existence, even after the tragic physical death presented in a cruel manner in the media:[12]

> We die as men. If you pay attention, we do not appear in the statistics as *travestis*, but as men. Not even the government recognizes our existence.

What you see is, "*o*" ["the" – male article in Portuguese] *travesti*, refer-ring to a male person, "*foi morto*" ["was killed" – male adjective in Portu-guese]. Our transition and the female condition are not respected. Now, see, I have a nephew. I always ask him to say "*a*" ("the" – female article in Portuguese) *travesti*. But then he reads the paper and sees "*o*" (male article in Portuguese) *travesti* and learns it this way, do you understand?

(Interview with Jasmim, on 27/08/2013, in Curitiba).

The denial of femininity by the family, mainly at the funeral, is a relevant ele-ment in the narrative about the construction of the social "enemy". The family mourning for the death of *travestis* and female transsexuals is reported by them with resentment, since the family many times does not respect their femininity, even after their death. The documents related to the death are written based on the male identity, which is considered an insult in face of their fight during the whole life for respect to their feminine condition, as reported by Violeta:

None of us wants to be buried as a man. I have a friend who had already separated the clothes she wanted to be buried in when she died, poor thing. It seemed she knew what would happen to her, she was killed. This is absurd, people are afraid of us even after we are dead, not even after death do we have peace. When the NGO prepare the funeral, everything is reg-istered using the female social name. The NGO even keeps a tomb where around twenty girls were buried because their families never reclaimed the bodies. But when the family is responsible for the funeral, they end up registering everything using the male name, so if you check the death cer-tificate, you'll find the male name. Therefore, our existence is erased.

(Interview with Violeta, on 19/09/2013, in Curitiba)

Transnecropolitics affect the Brazilian society on a daily basis in an organized manner to create the lives that are considered important as well as those that are disposable, such as *travestis* and transsexual women. The political, social and economic logic that operates the management of these people's lives is based on their exclusion while alive, the extermination of their bodies and the conceal-ment of their existence after their deaths.

Final Considerations

While we were writing this chapter, it was necessary to update the statistical data on *travestis*' and transsexuals' deaths since every 48 hours a trans life is violently ended in Brazil. Although the Brazilian trans activism is operating in Brazil, monitoring and denouncing the murder dynamics, these statistics are very high.

The text argued that these early and violent deaths suffered by the trans-people are part of a meticulous gear, a specific governance that targets *travestis* and female transsexual, who are young and live by the sexual market. The intersectional perspective of the group of victims helps to show a growing transnormativity that keeps inscribing the life experience of the trans communities in Brazil.

The narratives of people who took part in this chapter evidenced that the group is aware of the power dynamics that creates their vulnerability and murder. The widespread acceptance of the idea that *travestis* and female transsexuals as enemies against whom the society has to fight, makes their lives disposable and their deaths celebrated.

The transnecropolitics is specially operated via constant exclusion processes that result in the extermination of the body and the concealment of their existence after death. The understanding of trans death in Brazil as a political process might build up a path for the deconstruction of their lives as disposable lives. The catch phrase that starts this text "when a trans is killed, another thousand rise" has to be broadened in order to reach the Brazilian society as a whole so that they can shout: "when a trans is killed, we all rise"!

Notes

1 Trans woman and *travestis* are people who were assigned male at birth, but identify themselves as belonging to the female gender. In Brazil, transsexual identity is recent, and there is still tension between people who self-identify as *travestis* and trans woman in the LGBTI movement. For interviewed people, being a *travesti* implies keeping a penis and using it. But, above all, being a *travesti* is keeping a political identity of resistance, regardless of genitalia, as pointed out by Silva and Ornat (2016).

2 Map of the cases of murders of *Travestis* and Transsexuals in Brazil in 2017 reported by the Associação Nacional de Travestis e Transexuais (ANTRA) (National Association of *Travestis* and Transsexuals). Available at www.google.com/maps/d/viewer?mid=1yMKNg31SYjDAS0N-ZwH1jJ0apFQ&ll=-10.10973362929658%2C-20.126154816406256&z=3. Accessed on 09/08/2017.

3 All the participants are identified in the text with fictional names.

4 Another contradiction between desire and hate was raised by the survey presented in the porn site "Red Tube" which shows that Brazilians are the nationality that most search for films in which the main characters are trans women and *travestis*. At the same time, it is the country that kills the most *travestis* and transsexual people in the world. Source: https://goo.gl/LezGix. Accessed on 10/08/2017.

5 The action of the *Grupo Gay da Bahia* has to be highlighted. Since the early 1980s this group has presented statistics and systematization of murders of LGBTI people in Brazil by surveying news broadcast in the national media. Another important entity that produces data related to the murders in the global trans population is the non-governmental organization Transgender Europe.

6 It seems important to highlight that the news related to murders of transpeople shown in the media, which are ANTRA sources, might present distortions regarding the gender identities of the victims. The news stories do not include the victim's self-identification as transsexual women or *travestis*. They are categorized arbitrarily, by the common sense recognition that the society make of their bodies.

7 Data extracted from: http://blogs.odia.ig.com.br/lgbt/2015/01/29/dia-nacional-da-visibilidade-trans-e-celebrado-com-manifestacao-na-cinelandia-nesta-quinta-feira/. Accessed on 10 August 2017.

8 These body transformations may include injection of industrial silicone and taking hormones.

9 ANTRA claims that the entrance and maintenance of *travestis* and transsexuals in education institutions were made easier by the right to use their "social name", regulated at

Federal level by the Decree n. 8727 of 28 April 2016. However, several Brazilian States had anticipated the possibility to use a social name in education and health institutions. The social name is that which corresponds to the gender identity, but does not substitute the "civil name". The civil name is that used in the documents and that corresponds to the sex assigned at birth. The change of civil registers in Brazil only occurs through a legal process and judicial authorization.

10 In Brazil, transsexuality is seen as an illness and is listed in the *Manual Diagnóstico e Estatístico de Transtornos Mentais (DSM – APA)* (Mental Disturbance Diagnostic and Statistics Handbook) and the International Code of Diseases (ICD – WHO).

11 The idea of transnormativity is used in this chapter as the adoption of conservative practical and political agendas and practices by transpeople, which ends up creating identitary hierarchies within the trans communities.

12 The newspapers expose photos of transpeople who were killed, showing injured and torn bodies, without showing any respect for the victims or those who love them.

References

Agamben, Giorgio. 1998. *Homo Sacer: Sovereign Power and Bare Life*. Stanford, CA: Stanford University Press.

Antunes, Pedro Paulo Sammarco. 2013. *Travestis envelhecem?* São Paulo: Annablume.

Balzer, Carsten (Carla La Gata), Hutta, Jan Simon, Adrián, Tamara, Hyldal, Peter and Stryker, Susan. 2012. *Transrespect versus Transphobia Worldwide. A Comparative Review of the Human Rights Situation of Gender-variant/Trans People*. Berlin: Transgender Europe (TGEU).

Bardin, Laurence. 2004. *Análise de Conteúdo*. 3rd ed. Lisboa: Edições 70.

Berlant, Lauren. 2007. "Slow Death (Sovereignty, Obesity, Lateral Agency)". Critical *Inquiry* 33(4): 754–780.

Boulevard, Gláucia. 2013. "Vida de travesti é luta! Luta contra a morte, luta contra o preconceito, luta pela sobrevivência e luta por espaço". In *Geografias malditas: corpos, sexualidades e espaços*, edited by Joseli Maria Silva, Marcio Jose Ornat and Alides Baptista Chimin Junior, 69–81. Ponta Grossa: Todapalavra.

Brasil. 2013. *Relatório sobre Violência Homofóbica no Brasil: ano de 2012*. Brasília: Secretaria de Direitos Humanos da Presidência da República.

Butler, Judith. 1993. *Bodies that Matter: On the Discursive Limits of "Sex"*. London: Routledge.

Butler, Judith. 2004. *Precarious Life: The Powers of Mourning and Violence*. London and New York: Verso.

Cabral, Vinícius, Silva, Joseli Maria and Ornat, Marcio Jose. 2013. "Espaço e morte nas representações sociais de travestis". In *Geografias malditas: corpos, sexualidades e espaços*, edited by Joseli Maria Silva, Marcio Jose Ornat and Alides Baptista Chimin Junior, 246–275. Ponta Grossa: Todapalavra.

Collins, Patricia Hill. 2000. *Black Feminist Thought: Knowledge, Consciousness, and the Politics of Empowerment*. New York: Routledge.

Crenshaw, Kimberlé Williams. 1991. "Mapping the Margins: Intersectionality, Identity Politics, and Violence against Women of Color". *Stanford Law Review* 43(6): 1241–1299.

Davis, Kathy. 2009. "Intersectionality as Buzzword: A Sociology of Science Perspective on What Makes a Feminist Theory Successful". *Feminist Theory* 9(1): 67–85.

Doan, Petra L. 2010. "The Tyranny of Gendered Spaces – Reflections from Beyond the Gender Dichotomy". *Gender, Place & Culture: A Journal of Feminist Geography* 17(5): 635–654.

Foucault, Michel. 1978. *History of Sexuality*, vol. 1. New York: Vintage.

Foucault, Michel. 1995. *Discipline and Punish: The Birth of the Prison*. New York: Vintage Books.

Foucault, Michel. 1997. *Society Must Be Defended: Lectures at the Collège de France, 1975–1976.* New York: St. Martin's.

Grupo Gay da Bahia. 2014. *Assassinato de homossexuais no Brasil.* Relatório anual do Grupo Gay da Bahia. www.ggb.org.br/assassinatos%20de%20homossexuais%20no%20brasil%20 2011%20GGB.html (accessed March 12, 2017)

Haritaworn, Jin, Kuntsman, Adi and Posocco, Silvia. 2014. *Queer Necropolitics.* Abingdon: Routledge.

Hutta, Jan Simon. 2013. "Queer Necropolitics and other Power Geometries in Brazil". *Geographies of Sexualities Conference in Lisbon,* September (Session Presentation).

IBGE. 2013. *Pesquisa nacional por amostra de domicílios.* Rio de Janeiro: IBGE.

Johnson, Austin H. 2016. "Transnormativity: A New Concept and Its Validation through Documentary Film About Transgender Men". *Sociological Inquiry* 86(4): 465–491.

LeBlanc, Fred Joseph. 2010. "Unqueering Transgender? A Queer Geography of Transnormativity in Two on Line Communities". Master thesis – Gender & Women's Studies. Victoria University of Wellington. Wellington. p. 143.

Lee, Debora. 2013. "A geografia de uma travesti é uma barra, é matar um leão a cada dia". In *Geografias malditas: corpos, sexualidades e espaços,* edited by Joseli Maria Silva, Marcio Jose Ornat and Alides Baptista Chimin Junior, 27–38. Ponta Grossa: Todapalavra.

Maddrell, Avril and Sidaway, James D. 2010. *Deathscapes: Spaces for Death, Dying, Mourning and Remembrance.* Farnham: Ashgate Publishing Limited.

Mbembe, Achille. 2003. "Necropolitics". *Public Culture* 15(1): 11–40.

Mbembe, Achille. 2008. "Necropolitics." In *Foucault in an Age of Terror. Essays of Biopolitics and Defense of Society,* edited by Stephen Morton and Stephen Bygrave, 152–182. New York: Palgrave Macmillan.

McCall, Leslie. 2005. "The Complexity of Intersectionality". *Signs: Journal of Women, Culture and Society* 30(3): 1771–1800.

McDonald, Myfanwy. 2006. "An Other Space". *Journal of Lesbian Studies* 10(1–2): 201–214.

Nikaratty, Leandra. 2013. "O que mais me marcou na vida é ser barrada e não poder entrar nos lugares: esta é a geografia de uma travesti". In *Geografias malditas: corpos, sexualidades e espaços,* edited by Joseli Maria Silva, Marcio Jose Ornat and Alides Baptista Chimin Junior, 239–254. Ponta Grossa: Todapalavra.

Ornat, Marcio Jose. 2009. "Espacialidades travestis e a instituição dos territórios paradoxais". In *Geografias Subversivas: discursos sobre espaço, gênero e sexualidades,* edited by Joseli Maria Silva, 177–210. Ponta Grossa: Todapalavra.

Puar, Jasbir. 2007. *Terrorist Assemblages: Homonationalism in Queer Times.* Durham, NC: Duke University Press.

Silva, Joseli Maria and Ornat, Marcio Jose. 2016. "Transfeminism and Decolonial Thought: The Contribution of Brazilian Travestis". *TSQ: Transgender Studies Quarterly* 3(1–2): 220–227.

Stryker, Susan. 2014. "Biopolitics". *TSQ: Transgender Studies Quarterly* 1(1–2): 38–42.

Stryker, Susan, Currah, Paisley and Moore, Lisa Jean. 2008. Introduction: Trans-, Trans, or Transgender? *WSQ: Women's Studies Quarterly* 36(3–4): 11–22.

Index

Note: Page numbers in *italics* indicate a figure on the corresponding page.

Printed in the United States
by Baker & Taylor Publisher Services